Conflicts of Conscience in Health Care

Basic Bioethics
Glenn McGee and Arthur Caplan, editors

Conflicts of Conscience in Health Care

An Institutional Compromise

Holly Fernandez Lynch

The MIT Press
Cambridge, Massachusetts
London, England

For information about special quantity discounts, please e-mail special_sales@mitpress.mit.edu

This book was set in Sabon on 3B2 by Asco Typesetters, Hong Kong.
Printed on recycled paper and bound in the United States of America.

Library of Congress Cataloging-in-Publication Data

Lynch, Holly Fernandez.
Conflicts of conscience in health care : an institutional compromise / by Holly Fernandez Lynch.
    p.   ;   cm. — (Basic bioethics)
Includes bibliographical references and index.
ISBN 978-0-262-12305-1 (hardcover : alk. paper)
1. Refusal to treat—Moral and ethical aspects—United States. 2. Physician and patient—Moral and ethical aspects—United States. 3. Physicians—Professional ethics—United States. 4. Medical laws and legislation—United States.
5. Conscience—United States. 6. Medical ethics—United States. I. Title.
II. Series.
[DNLM: 1. Ethics, Medical—United States. 2. Refusal to Treat—ethics—United States. 3. Conscience—United States. 4. Health Facilities—United States.
5. Health Services Accessibility—United States. 6. Refusal to Treat—legislation & jurisprudence—United States. W 50 L987c   2008]
R727.36.L96   2008
174.2—dc22   2008005529

10   9   8   7   6   5   4   3   2   1

For Dad, who taught me how to write, painful as it was,
And for Bill, the most patient and devoted husband I could ask for.

Contents

Contents

Series Foreword

We are pleased to present the twenty-fourth book in the series Basic
Bioethics. The series presents innovative works in bioethics to a broad
audience and introduces seminal scholarly manuscripts, state-of-the-art
reference works, and textbooks. Such broad areas as the philosophy of
medicine, advancing genetics and biotechnology, end-of-life care, health
and social policy, and the empirical study of biomedical life are engaged.
Glenn McGee
Arthur Caplan

Basic Bioethics Series

Preface

Often, when one seeks a compromise and attempts to please those standing on both sides of a divisive issue, she ends up angering everyone and pleasing no one. That is precisely what presidential hopeful Rudy Giuliani seemed to do when he stated at a 2007 Republican debate that "it would be okay to repeal [*Roe v. Wade*]. It would be okay also if a strict constructionist judge viewed it as precedent, and I think a judge has to make that decision." This thread-the-needle response angered both pro-choice and pro-life groups, and I fear that my attempt at a compromise for the problem of conflicts of conscience in health care will have a similar fate. Those opposed to conscientious refusal by physicians are likely to charge me with not going far enough to protect patients, and those who support a physician's claim to engage in refusal may allege that I have not been sufficiently sympathetic to their effort to avoid complicity in immoral acts.

In a compromise, neither side obtains its ideal solution—that is the price we pay to avoid a stark, dichotomous winner-loser outcome—but that does not mean that both parties are losers. Instead (at the risk of sounding trite), we are all winners, particularly if we can simultaneously enable patients to obtain the medical services they need and want while preserving the medical profession as a group of conscientious individuals who are unwilling to do anything and everything they are capable of. To see this, I urge you to set aside your usual sympathies, be they for patients requesting medical services that they can only obtain from a physician or for physicians who are asked to do things that they believe to be wrong. Try earnestly to take on the opposite perspective, or at least a neutral one, while reading this book.

This open-minded philosophy has helped me tremendously throughout the long process of thinking about the problems raised by conflicts of conscience in health care and I have changed my mind on more than one occasion. Be assured that I have not sought a compromise just for the sake of it, though the reality is that progress in resolving these conflicts will be impossible if we continue to adopt a winner-take-all strategy. Rather, I have become genuinely convinced that neither patients' nor physicians' claims clearly trump those of their opponent; compromise is right not just for pragmatic reasons, but for more abstract normative ones as well. Therefore, it is essential for everyone to be willing to engage in an ideological shift, though from the perspective of conscience clauses currently on the books, patients have everything to gain and physicians everything to lose.

This entreaty aside, a few additional notes to the reader are in order. First, I must make it clear from the outset that this is not a work of moral philosophy that grapples with the fundamental nature of conscience, tries to discern where it comes from, or attempts to justify the conclusions drawn in its name. Instead, this is a book with a legal trajectory, which takes as a starting point the fact that conscience is an incredibly powerful force in our lives and recognizes that this force has created some practical problems in the context of modern health care. The goal is to resolve these problems in as comprehensive a manner as possible, while leaving the philosophy to the philosophers, who can do it far better than I.

Additionally, I must provide a brief disclaimer regarding scope. The issues surrounding pharmacist refusals are interesting in their own right as a unique aspect of the professional conscience debate. This is particularly true in light of the criticism that refusing pharmacists are interfering in the doctor-patient relationship, as well as the fact that forcing pharmacists to fill prescriptions in violation of their conscience is not likely to pose health risks to patients, unlike, for example, forcing a doctor to perform a complicated surgical procedure against her will. Nevertheless, this book will focus almost exclusively on physicians because their professional obligations may be quite different from those held by other health-care workers, who are widely understood to have less professional autonomy. More important, these nonphysicians are generally employed by some overarching institution, such as commercial pharmacies, that can bear the burden of referral or can provide controversial services when individuals refuse to do so. The frequent absence of these employment relationships is a unique characteristic of the conscience clause debate as it applies to physicians, which makes it more difficult to resolve and thus calls for a discussion separate from that regarding other health-care professionals. For these reasons, pharmacist refusals are introduced in the book solely for instrumental purposes, when the debate surrounding them has something important to offer on a broader scale.

Similarly, the book will largely avoid discussion of institutional conscience, such as moral objections raised by religious medical centers, despite the fact that this issue has also been heavily debated. This purposeful decision to limit the scope of the present inquiry is based on the desire to look carefully at the interaction of the professional obligations and moral beliefs of individual persons, as well as the fact that there is greater support, comparatively speaking, for individual conscience, which potentially indicates greater room for acceptance of the type of compromise solution suggested here. Readers interested in these other instantiations of the conscience clause dilemma may be able to draw important analogies from the arguments and conclusions presented in this book, and are of course encouraged to continue on, but should be aware of its primary focus on refusals by individual doctors.

In terms of the number of people affected, either by physician refusals or the absence of conscience clause protection, the issue of personal morality in the medical sphere may seem rather trivial. The vast majority of patients will never be subjected to a physician's refusal, and if they are, they will be able to access the desired service elsewhere with relative ease. Similarly, the vast majority of physicians unwilling to provide particular services simply choose different specialties of which those services are not a part or find other willing physicians to cover for them. Certainly, this problem is of lesser magnitude than other grave concerns facing bioethicists today, such as the vast number of uninsured Americans, extraordinary global health disparities, and issues of justice in the allocation of scarce resources.

So why write a book about it? Because while it may be true that the conscience clause issue has received more than its fair share of public attention, it is nevertheless worthy of serious debate given the significant moral issues it raises. The matter is certainly important to the minority of patients for whom a physician's refusal results in complete unavailability of the medical service sought, and the number of patients affected is likely to grow as technology moves quickly into territory that many find ethically murky. The issue is also paramount to physicians who strongly value their personal moral autonomy even in the professional context. Conscience is vital to individual integrity, so calls to quash it must be taken seriously, as should denials of access. On a more abstract level, the conscience clause debate is deeply connected to broader questions of professionalism and the role of physicians as moral arbiters, protecting society from technology run amok. Thus, there are pressing, challenging, and interesting questions to be answered about which I have something new and (I think) useful to say—that is all the motivation this author needs.

Acknowledgments

This project was undertaken while I was a member of the inaugural cohort of academic fellows at the Petrie-Flom Center for Health Law Policy, Biotechnology, and Bioethics at Harvard Law School. The Center aims to promote and support interdisciplinary scholarly research into important issues at the intersection of law and health policy and is committed to the neutral exploration of new ideas.

First, I must thank the Center's faculty director, Einer Elhauge, who encouraged me to pursue this idea and give it book-length treatment, offered insightful commentary on my arguments along the way, and patiently advised me throughout the publication process.

I am also hugely indebted to Glenn Cohen, my "fellow fellow," who offered incredibly helpful advice on this project since day one. Glenn is the epitome of what a scholar should be: intellectually curious, dedicated to getting things right, and extraordinarily generous with both his time and encouragement. This book would have suffered tremendously without him.

The arguments in this book were also shaped and improved by the keen and discerning comments of anonymous peer reviewers, as well as through conversations with Norman Daniels, Michelle Mello, Ben Roin, and Talha Syed. Additionally, I am thankful to participants in the Center's opening workshop, particularly Dan Brock and Frances Kamm, whose comments at that early stage stayed with me throughout the writing process, as well as to those who were willing to discuss my ideas over the dinners that followed the fall 2006 Health Law Policy and Bioethics workshops.

I am especially thankful to Hermes Fernandez, "a fine father and a fine lawyer," for his assistance in reviewing and offering suggestions on the entire manuscript—all on an extremely tight deadline. He also did his absolute best to help me with writers' angst, for which I am grateful beyond words. Thank you also to my incredible mother, husband, sister, brother, and friends for their unconditional support, encouragement, and attempts to keep me sane—even my loyal writing assistants/ distractions, Bubba and Isis, did their part. I truly could not have done this without all of your help and I hope you know how deeply my appreciation runs.

Finally, thanks are owed to Kathy Paras, the Petrie-Flom Center's tireless Program Administrator, for her words of encouragement and assistance with a wide variety of logistical matters, and to the fine editors at MIT Press.

The views expressed are solely those of the author and should not be attributed to Hogan & Hartson, LLP or its clients,

Introduction

"When the physician becomes merely a biomedical technician whose role is never in a moral sense 'to question why,' medicine enters dangerous ground."[1]

"The door to 'value-driven' medicine is a door to a Pandora's box of idiosyncratic, bigoted, discriminatory medicine."[2]

Over a decade ago, outside the heated context of the current debate surrounding professional refusal clauses, Pope John Paul II made a powerful statement about the nature of conscience, complicity in morally objectionable actions, and avoidance of injustice, which can be generally accepted by both the religious and nonreligious alike, located at nearly every point along the political spectrum. He opined that "to refuse to take part in committing an injustice is not only a moral duty, it is also a basic human right. Were this not so, the human person would be forced to perform an action intrinsically incompatible with human dignity, and in this way human freedom itself, the authentic meaning and purpose of which are found in its orientation to the true and the good, would be radically compromised."[3] The most compelling question raised by professional refusal legislation, however, is whether this basic human right identified by John Paul II is one that is or ought to be at least partially waived by the decision to enter a particular profession.

Refusal statutes have been around for decades. They generally protect health-care professionals—mostly physicians, although they are increasingly being expanded to cover other members of the health-care team—from consequences that might otherwise arise from their refusal to provide certain medical services to patients solely because those services are incompatible with the professional's personal moral beliefs. However, proposals have also begun to sprout up that would require medical professionals to provide even those services they find morally objectionable on pain of job loss or liability unless they can offer a valid medical reason for refusing to comply with a patient's request.

Though there is nothing inherently new about the access concerns these laws raise, they have only recently returned to the forefront of public consciousness as refusals by pharmacists to fill valid prescriptions for emergency contraceptives have been brought to light through a handful of emotional media reports. These stories chronicle humiliated women being forced to spend hours searching for someone to fill their prescription, wasting precious time in which the drugs are most effective. In the interest of journalistic objectivity, however, these stories also tend to offer sympathy to the dramatic foe, health-care professionals in danger of losing their jobs for simply trying to do what they think is right. The reports have sparked renewed discussion about the professional obligations of all types of health-care providers to comply with patient requests for all types of medical services, ranging far beyond the limited scope of contraception and abortion, although those services have received the brunt of public attention.

Lest we minimize this sort of conflict as being overly hyped by the media, consider the fact that an empirical study reported in the *New England Journal of Medicine* in 2007 suggests that as many as forty million Americans—one in seven—may be cared for by physicians who are undecided or believe they have no obligation to disclose information about medically available treatments that they consider personally objectionable. As many as one hundred million Americans—one in three—may be cared for by physicians who are undecided or believe they have no obligation to provide patients with referrals to willing providers when conflicts of conscience occur.[4] In a less generalizable survey conducted online by *Self* magazine, nearly one in twenty respondents reported that their doctors had refused to treat them on some occasion for moral, ethical, or religious reasons.[5] The following examples, based on composites of real cases, demonstrate the reality of what these numbers might mean in practice and show that conflicts of conscience are an issue not only for women of childbearing age, but can also affect the young and old, men and women, religious and nonreligious alike. As you read them, think about your own career—what would you do if asked to perform a task to which you were genuinely morally opposed and what consequences would you likely face? Should that be the case?

Fiona McLoughlin, age sixty-three, is a lifelong resident of Oregon, where physician-assisted suicide was legalized by the Death with Dignity Act in 1994.[6] This statute authorizes physicians to expressly prescribe a lethal dose of medication that their patients can self-administer to end their own lives. The act grants legal protection to physicians providing such prescriptions to patients over the age of eighteen who are capable of making and communicating health-care decisions on their own, who have been diagnosed with a terminal illness predicted to result in death within six months, and whose competency, diagnosis, and prognosis have been confirmed by a second physician. Importantly, the statute does not require physicians to provide this service to their patients and protects doctors from "censure, discipline, suspension, loss of license, loss of privileges, loss of membership or other penalty" for refusing to write a lethal prescription.[7]

McLoughlin has recently been diagnosed with terminal cancer. Her oncologist, Dr. Ryan Bradley, tells her she has only three months to live, and will likely experience extraordinary pain, even with the best palliative care available. Until this diagnosis, McLoughlin has been quite healthy, both physically and mentally. However, she has a family history of cancer, and watched her own mother and sister die agonizing deaths from the very disease with which she is now suffering. Their deaths were difficult not only for them, but also for their families and friends, who, while thankful for every additional day they were able to spend with their loved one, quickly came to understand that there are some fates worse than death.

McLoughlin asks Dr. Bradley to write her a prescription under the act, telling him that she might never use it, but explaining that knowing she could end her suffering when it became unbearable would ease her mind tremendously, allowing her to enjoy her final days to the greatest extent possible. Expressing deep regret, Dr. Bradley refuses her request. He agrees that McLoughlin satisfies all of the legal criteria required by the statute, but tells her that while he wishes he could help, it would be a violation of his professional oath to do no harm and the American Medical Association's Code of Ethics if he were to assist in ending her life in any way.[8] He will continue to make her as comfortable as possible, but that is all he can do.[9]

Is Dr. Bradley's refusal appropriate, notwithstanding the fact that it is legally protected by the Death with Dignity Act? What if his refusal were instead based on a religious belief that ending one's own life is sinful and that God will not give anyone more than they can handle? Does he have an obligation at least to help McLoughlin find a doctor who is willing to write the prescription, and what if such a physician turns out to be quite difficult to find? Is the propriety of Dr. Bradley's refusal contingent on the reasonable availability of the legal medical service from another source? As these added layers of complexity demonstrate, there are many grounds for disagreement when it comes to regulating the expression of conscience in a professional setting and the answers are far from clear. With these questions in mind, consider another recent controversy.

Jack Simione is a pediatrician in a small Mississippi town who has been practicing for thirty years. He is well-respected in his community and widely loved by his patients. He is also a minister at the local Protestant church, attended by the vast majority of his patients' families, and is active in organizations that promote "no sex is safe sex" and "wait until marriage" educational programs for adolescents. He has many posters and pamphlets with these messages around his office, as well as posters with inspirational religious adages, leaving no question as to where he stands and no surprises for those seeking his care. Dr. Simione frequently discusses abstinence with his teenage patients, on both religious and safety grounds, and he also discusses ways for parents to reinforce these messages to their children.

Recently, the parents of several of Dr. Simione's patients have begun to request that their daughters be vaccinated against human papillomavirus (HPV), a sexually transmitted disease that has been shown to play a role in certain types of cervical cancer. While generally supportive of Dr. Simione's abstinence campaigns, the parents are aware that teenagers and young adults will often not be convinced and may bow to immaturity, peer pressure, or simple curiosity. Further, while they hope that their children will wait for marriage before they engage in sexual intercourse, they are aware that not all marriages are monogamous and that the risk of HPV and other sexually transmitted diseases will not disappear entirely. They want to do everything they can to protect their daughters from HPV and cervical cancer and have been convinced by various news stories and advertisements that the benefits of the newly available vaccine outweigh its risks. Dr. Simione is torn between providing appropriate preventative care to his patients and contradicting his own personal beliefs, fearing, logically or not, that protection against an STD will encourage premarital sex, promiscuity, and a generally un-Christian lifestyle.[10]

Suppose Dr. Simione ultimately decides that his religious beliefs preclude the provision of the vaccine. Has he shirked his obligations as a physician, inappropriately intruding on the personal choices of parents with regard to how to promote the best interests of their children? Does it matter if the vaccine is easily attainable elsewhere? On a broader level, were Dr. Simione's abstinence-only messages appropriate and would it matter if some parents chose him as a pediatrician based on their agreement with that message, or that parents at least had the benefit of advanced notice of his moral views? What if his refusal were grounded in concern that the vaccine simply had not yet undergone sufficient testing to provide adequate evidence of its safety and efficacy?[11]

Again, the issues raised by this scenario are incredibly complex. While some observers will believe Dr. Simione clearly did the right thing and others will believe he clearly violated his responsibilities, still others will stand torn in the middle, unsure how the conflict between doctor and patient autonomy (or here, substituted parental authority) ought to be settled. Now consider one final example.

Sara and Paul Lazar approach Dr. Lucy Gray, a fertility specialist in Rhode Island recommended by a family friend, about the possibility of conceiving a "savior sibling" for their son, Eric.[12] Eric has been diagnosed with a rare inherited disease called Fanconi anemia and will shortly be in need of a bone marrow transplant, but neither Sara nor Paul is a suitable donor, nor are any of their close relatives. They hope that they will be able to use in vitro fertilization (IVF) to create several embryos, selectively implanting those that would be certain marrow matches. They believe that this is the best option since they might not conceive a suitable donor child through random natural conception soon enough to save Eric; they cannot wait through multiple pregnancies resulting in nonmatching children and would not abort a nonmatching fetus in order to try again. Use of IVF and preimplantation genetic testing, on the other hand, could ensure that a pregnancy would result in a suitable donor, risk of miscarriage aside. However, when the Lazars explain their situation, Dr. Gray informs them that she cannot help. She is terribly sorry for Eric's condition, but she thinks it is wrong to conceive a child just to be used in this way.[13]

Again, is Dr. Gray's refusal appropriate? If she were otherwise willing to selectively implant embryos with or without certain characteristics, such as only male embryos or only those lacking the gene for Tay-Sachs, would her refusal now be inconsistent—and does that matter? At a broader level, does Dr. Gray have any right to inquire into the suitability of her patients to become parents or to consider their reasons for having a child? Does she have any special professional or moral obligations given that her practice involves the introduction of new lives into the world? Perhaps the sort of dissent she has expressed plays a valuable role in questioning the propriety of new medical technologies and how they ought to be used, but perhaps she has overstepped her bounds in denying a service that is legal and available only through the medical profession.

As these examples demonstrate, patients will often want something that their doctors are unwilling to provide for a variety of reasons. Opinions about the appropriate outcomes of such conflicts will depend on the precise factual situation, such as the level of the patient's need for the service in question, the extent of damage to the physician's personal integrity if forced to violate his conscience, and whether the physician's refusal is based on his religious beliefs, secular moral beliefs, discriminatory attitudes, expert medical opinions, or even shrewd business choices about which services will be reimbursed at the highest rates. We might also consider whether the physician at least made the patient aware of all her possible options even if the physician would refuse involvement with some of them. The physician's refusal generally appears least problematic in nonemergency situations where she is prepared to refer the patient to a willing provider, and when that provider is located within a reasonable distance.

Each of these factors plays a role in whether and to what extent observers believe that physicians should be forced to do things they genuinely consider wrong, and in defining those things doctors ought to be willing to do if they desire the privilege of membership in the medical profession. Unsurprisingly, they also play a role in the scope and limitations of legislation governing the refusal of health-care professionals to comply with a variety of patient requests. These laws range from the very broad, allowing all health-care workers and health-care entities to avoid negative consequences associated with their refusal to perform, cover, or be involved in any way with services they find objectionable, to the very narrow, protecting only physicians who refuse to perform abortion or sterilization services in nonemergency circumstances.

Both those in favor of and opposed to refusal statutes have adopted the colloquial phrase "conscience clause" to describe them. Proponents seem to have adopted the term for its evocative value; after all, since childhood, many of us have been taught to let our conscience be our guide and it can be extraordinarily troublesome to think of situations in which our ability to do so might be interfered with by a government mandate based solely on our career choice. Opponents, on the other hand, tend to use the phrase pejoratively, suggesting that something as weak as mere personal beliefs should not be permitted to interfere with what they consider to be rights of access to various medical interventions. Both sides are arguing in favor of freedom, differing only in the priority they are willing to give various freedoms when they come into conflict. The hard fact at the center of the conscience clause controversy, however, is that any solution will involve the restriction of someone's liberty to at least some degree, and the best approaches involve sacrifices all around.

At one extreme, commentators claim an unfettered right to conscientious refusal, arguing that physicians and patients are both human persons and thus have equivalent claims to personal moral autonomy, professional status notwithstanding. Further, they claim, the fiduciary relationship between doctors and patients does not mandate that physicians must always bow to the demands of patients, particularly since similar fiduciary obligations of parents, spouses, and attorneys do not preclude these individuals from refusing assistance to their children, spouses, and clients seeking inappropriate ends. They emphatically stress that "there is a difference between service and servitude."[14]

At the other extreme are those concerned about what conscientious refusal will mean for patients. Some of these commentators view patient autonomy as imposing a duty on physicians to provide any legal medical procedure a patient can dream of. More reasoned patient advocates focus on the potentially devastating impact widespread physician refusal based on conscience will have on patient access, particularly in rural and underserved areas. They argue that health-care workers should not be permitted to impose their moral views on patients, and point out that these workers willingly enter their field and therefore must adopt its corresponding obligations.

At this point in the debate, the standard arguments on either side have become familiar, and can be summarized with relative ease. John K. Davis has offered an excellent nutshell version that avoids caricature:

Some ethicists say the doctor has a duty to at least refer and facilitate a transfer to another doctor, and probably to counsel the patient as well. Others argue that the doctor should not refer the patient, or at least not when the treatment the patient wants is sufficiently immoral. It can also be argued that the doctor should perform the procedure anyway, if there is no other doctor available to do the procedure in time. Some say the doctor should reveal his moral views to the patient, but it is not clear that doctors should provide uninvited moral advice. Some writers advise doctors to make their moral views known up front, but that may require a degree of foresight it is not always reasonable to expect. Some go so far as to recommend that doctors stay out of a specialty altogether when their moral values preclude them from performing procedures considered standard in that specialty. Other writers deny this, arguing that medicine loses when moral dissenters are excluded.[15]

The dichotomies are clear and both sides claim that accepting the position of their opponent would introduce grave dangers into the medical arena—ushering in either an era of permissible discrimination and balkanized medicine or an era of medicine run amok, bereft of any shred of moral restraint. They are each right that embracing the other extreme would be inappropriate at best and perilous at worst, but as with most arguments at the edges, neither end of the spectrum is sufficiently nuanced to appropriately deal with conflicts of conscience in health care. Unfortunately, a simple solution is nowhere in sight, as both patients and doctors advance strong arguments in support of their claims that are not easily swept aside.

In response to this inherent difficulty, many commentators offer proposals that are so vague as to be essentially meaningless. For example, Thomas Beauchamp and James Childress, two of our era's most influential bioethicists, attempt a middle-of-the-road approach, stating that "if a physician wishes to withdraw, his or her conscientious convictions should be respected, and he or she should be free to withdraw —assuming that the requested actions are not among the general responsibilities of physicians."[16] However, without defining which responsibilities are so integral to a physician's role that they cannot be avoided on grounds of conscience, a major challenge, Beauchamp and Childress's caveat risks swallowing their rule. While on the right track, they do little more than throw up their hands with regard to the fundamental issue of patient access, asking questions, but not answering them: "Even if the physician does have a right to protect his integrity, does this right outstrip or trump the rights of others? Can a physician justifiably say 'I have a right of conscience' and use this trump to escape the clutches of a moral dilemma?"[17]

Beauchamp and Childress should not be faulted for their vagueness, given that their statements were offered in the context of a much broader volume the goal of which was to lay out a variety of principles to guide bioethical decision making. They were not endeavoring to resolve conflicts of conscience specifically, but instead were simply using them as an example, and so for that purpose, simply highlighting the issues was sufficient. Unfortunately, however, many discussions that do directly target the issue of conscientious refusal similarly resort to mere lists of rhetorical questions, or state that a compromise solution is necessary, without offering any specifics with regard to what such a compromise might look like.[18] These failures offer an important commentary on the seeming intractability of this moral quandary, but it is clear that posing rhetorical questions will do little to further the conscience clause debate, which demands to be addressed head on. The goal of this book is to do just that, finding a way to resolve the problems associated with conflicts of conscience for both doctors and patients through a concrete and detailed proposal for change.

As a society, we should not expect or demand that physicians always check their personal religious and moral beliefs at the door, since that is both impossible and unnecessary, as the most reasonable commentators on this problem acknowledge. Instead, we should attempt to recognize a place for personal morality whenever possible. If we treat physicians merely as sophisticated medical vending machines,[19] we risk further damage to medical professionalism and the doctor-patient relationship, which are already under attack by everything from managed care to direct-to-consumer advertising. Aside from their medical training, physicians also have a wide variety of life experience, and the best doctors tend to bring this nonclinical experience to bear when dealing with patients. Arguably, patients would lose a great deal if we forced doctors to completely abandon their own morality in the clinical setting—and physicians most certainly would. Nevertheless, we must hold physicians to the promises they make in their professional practices and must somehow account for the vulnerability of patients who depend on them for care.

The professional role of physicians can best be understood in the context of their state-granted licenses, conferring on the profession a collective monopolistic power over the provision of most medical services. Because of this gatekeeping role, under which patients cannot gain access to desired services without a physician first opening the gate, many commentators have argued that physicians have a professional obligation to subordinate their personal moral beliefs to patient requests. However, simply because *a* physician is needed to open the gate does not mean that *all* physicians must do so, allowing a distinction between obligations of the profession as a whole and those borne by its individual members—a bifurcated model of medical professionalism in which the whole is recognized as greater than the sum of its parts.

This book draws on that important distinction to advocate an institutional solution that maintains the availability of desired medical services while preserving the personal moral integrity of physicians in most circumstances. Recognizing the relative success of institutional solutions to the conscience dilemma in both the pharmacy and hospital contexts, a similar compromise is available on a broader scale if state licensing boards bear the responsibility of ensuring access to a sufficient supply of physicians willing to open the gate for patient requests. Rather than the all-or-nothing proposals that have been encouraged by both patients' rights and physicians' rights advocates, under this approach, physicians with a conscience will not be excluded from their chosen specialty, patients will not be forced to bear the brunt of beliefs that differ from their own, and there will be room for physicians and patients to create relationships based on deeply held, shared moral values. Thus, revising the responsibilities of licensing boards, as well as asking for appropriate concessions from both doctors and patients, will create a workable and balanced, albeit controversial and imperfect, resolution to the ongoing conscience clause debate.

Of course, imposing this new burden on licensing boards will be complicated and contentious, especially in the eyes of those who see this solution as offering far too much accommodation to objecting physicians. It will require a fundamental restructuring of licensing boards and their responsibilities, including the development of new levels of expertise, as well as a significant investment of resources. Unless they undertake serious changes, licensing boards as they currently exist could not successfully implement the strategies explored throughout this book for mediating between concerns for physician conscience and the availability of medical services—but change they must. The conscience clause dilemma has become too serious for these professional boards to ignore. Criticism is certainly expected, but the following chapters will show the failings of other potential solutions and the importance of a detailed compromise.

While this book strives to offer a solution that is both practical and normatively correct, it is important to stress that for the purposes of this project, getting the right answer to the question of what should be done about conflicts of conscience in health care trumps concerns about the political feasibility of the ultimate proposal. There is no doubt that opening state medical practice statutes and imposing these tremendous new obligations on licensing boards could hit a "third rail" in American politics and would likely be received with political combativeness. However, our options seem to be: (1) to force physicians to do things they wish to avoid, which certainly would not be welcomed by many; (2) to allow physicians to refuse their services on moral grounds while ignoring the potential access problems posed for patients, which is the present approach in most states, but is also quite troublesome; or (3) to come up with some middle ground. Only an institutional solution seems capable of accommodating both doctors and patients, and this book goes to great lengths to explain why accommodating only one or the other is inappropriate.

This is not a project about political advocacy, and so it is not confined to solutions that would be politically acceptable at this time. As far as conscientious refusal in medicine goes, all options are controversial, so that should not be used as a criterion for dismissing one or another. Instead, the *relative* degree of feasibility compared to other options must be the question. Further, what is politically infeasible in one state or country might be feasible in others, and what is not feasible now might become feasible later. Indeed, academic endeavors such as this one can, as time passes, change what becomes politically acceptable by exposing the lack of any justifiable foundation for the status quo. It is a well-established element of political theory that over time, "for some kinds of solutions, attitudes soften and ideas that were once anathemas gain support and can be adopted.... Promoters talk about their proposal and make people more familiar with it. They may also change it enough to make it more acceptable. As a result, ideas that failed the appropriateness test when first introduced tend to be introduced time and again, becoming better understood, more familiar, and easier to accept as time goes by."[20] With these considerations in mind, it becomes clear that we must not be bound by the current political winds. Instead, normative correctness should be our primary guide, and that criterion is satisfied.

The licensing board proposal is the right answer and it can be practicably implemented with sufficient political will. If it remains unpopular with powerful political interests, that would be unfortunate, but it would certainly not render this project worthless. If it turns out that this approach is unworkable or otherwise unacceptable, we will see that it is likely that no institutional solution is available to rescue physicians from the conscience clause dilemma—and if no institutional solution is feasible, the likelihood of attaining a compromise is dramatically reduced. Instead of asking what a compromise should look like, we will instead be faced with the question of whether refusing physicians or requesting patients have the stronger moral claim. The debate would at least move forward through the elimination of one original option and policymakers will be left to choose between less palatable extremes. This book argues, however, that there is no reason to go so far—at least not yet.

Part I offers a brief introduction to conscience clause legislation, exploring how it came into existence in this country just over three decades ago and why it has recently moved to the forefront of public attention after a relatively long dormancy. More important, it explains just what it is these statutes are doing for physicians. It examines the current conscience clause debate, noting the many shortcomings that have inhibited the creation of a workable solution, and then it moves on to the crux of the issue: professionalism.

Clearly, it would be inappropriate to allow physicians to refuse to satisfy obligations that they have accepted by virtue of entering the profession, but precisely what professional obligations do physicians bear? This section analyzes and critiques various models of medical professionalism, examining the descriptive consent paradigm, as well as patient-focused versus physician-centric models. It ultimately concludes that while the profession as a whole bears important gatekeeping responsibilities to ensure that the services entrusted to it through a collective legal monopoly are made available to patients, individual physicians have far wider latitude to negotiate individual promises with those seeking their care.

Part II explores the impact this model of professionalism has on the conscience clause debate. To bolster the personal integrity claims often made by refusing physicians, it begins with an account of why preserving physician conscience is good for society more broadly. It then proposes that encouraging doctor-patient matching based on the moral beliefs important to each party, a model advocated by Robert Veatch, would improve the clinical encounter and, as matching increased, avoid many conflicts of conscience altogether. The success of this model depends fundamentally on moral diversity within the profession, such that patients have a sufficient number of "matching" physician counterparts from which to choose. This is primarily an issue of access and availability that can be resolved if state licensing boards ensure that patient demand for various medical services is met by an adequate supply of morally willing physicians located within an appropriate geographic area.

Part III fleshes out this proposal by exploring which medical services licensing boards should have an obligation to ensure access to, describing what sort of access is acceptable, and offering a variety of examples that boards will be able to draw on in meeting their new responsibilities. However, if access problems remain as a result of conscientious refusals, patients should seek compensation from these boards rather than from refusing physicians. Physicians should not be subject to individual liability or professional discipline for their refusals even when they stand as the last doctor in town. At the same time, they should not have the benefit of protection from consequences when their refusals are based on unacceptable grounds, such as outright discrimination on the basis of sexual orientation or race, or if they fail to satisfy a number of other prerequisites.

The book concludes with a model statute that implements the institutional solution to the conscientious refusal problem proposed herein and stands in stark contrast to much of the current legislation concerning this matter. It is essential to note, however, that the licensing board mechanism is not a substitute for conscience clauses. They will still be necessary in some form to protect physicians and preserve moral diversity, but the institutional solution is a supplement designed to ensure access to a broad range of services that might otherwise be inhibited.

In many ways, conscience clauses have gone too far, but in some ways, they have not gone far enough. In a country that regards the broad protection both of personal moral beliefs and personal medical choices as central to our liberty, we cannot be content with simply sacrificing one freedom in favor of another when they come into conflict without first exploring more moderate alternatives. As Pope John Paul II recognized, the ability to act in accord with one's conscience is central to human dignity; on the other hand, actions can appropriately be prohibited when they interfere with the freedoms of others. This book argues, however, that there is a way to prevent the action of physician refusal from interfering in any significant way with patient interests. As the least restrictive alternative, creative institutional solutions to the conscience clause dilemma as it applies to physicians must be given serious consideration, as they are in the chapters that follow.

I

Conscience Clauses and Professionalism

1

A Primer on Conscience Clauses

For obvious reasons, conscience plays a major role in ethical endeavors, and those within the realm of bioethics are certainly no exception. Conscience frequently rears its head in bioethical decision making since these choices often raise deep questions about life and death, God and science, tradition and technology. Further complicating matters is the fact that doctors and their patients may hold widely disparate perspectives on these issues, but only one course of action can be chosen, which often cannot be taken back and which may have lasting consequences for either or both parties. For these reasons, some find it hard to believe that a physician's conscience is formally protected at all, while others are surprised to learn that this was not the case until about the time of the Supreme Court's landmark decision in *Roe v. Wade*. Then, the first refusal clauses began to make their way into responsive legislation.

A Brief History

Prior to that point in history, and even in response to a proper reading of *Roe*, there was no need for statutory protection of physicians. The law generally removed the sorts of choices that could have raised conflicts of conscience in the first place by banning controversial services like abortion and birth control. Additionally, since physicians generally wielded so much power in the doctor-patient relationship, conflicts of conscience were highly unlikely—or were at least unlikely to come to the attention of patients. Nevertheless, the recognition of a constitutional right against unduly burdensome state interference in a woman's personal decision about whether to carry her pregnancy to term sparked fear among those opposed to abortion that this negative right would soon be transformed into a positive one, by which physicians and other health-care workers could be forced to provide these services against their moral objection.

This fear led Congress to pass the Church Amendment in 1973,[1] its ironically religious title having nothing to do with its content, but instead referring to the statute's sponsor, Senator Frank Church. This legislation, still in effect today, was a direct response not to *Roe*, but rather to a district court's injunction requiring a Catholic hospital to allow sterilizations to be performed on its premises over its religious objection.[2] The injunction was promptly dissolved in the wake of the statute's protection of institutions and individuals receiving federal funds from being forced to provide abortion or sterilization services if those services would be contrary to their "religious beliefs or moral convictions."[3] The statute also forbids institutions receiving federal funds from discriminating against health-care employees based on their religious or moral convictions regarding these procedures.[4] The vast majority of states quickly followed suit, passing a flurry of conscience clause legislation of their own, filling in the gaps and offering protection not dependent on any particular funding source.[5]

Of course, the negative right established in *Roe* has not evolved into a positive right, and in fact seems to be moving in the opposite direction. Yet reactive conscience clauses remain on the books and have even expanded. Today, forty-six states, the District of Columbia, and the federal government have statutes protecting physicians from an incredibly broad array of consequences for refusal to participate in abortion procedures,[6] and Illinois,[7] Mississippi,[8] and Washington[9] allow refusal of *any* medical service to which the physician is morally opposed. Several other states extend their conscience clause protection to contraceptive and sterilization services, assisted reproduction, human cloning, fetal experimentation, physician-assisted suicide, and withholding or withdrawing life-sustaining treatment. Further, many of these statutes reach beyond physicians to cover other health-care providers, such as nurses and pharmacists, as well as health-care institutions, such as hospitals and insurers.[10]

In addition to *Roe v. Wade*'s integral impact on the addition of conscience clauses to our legal landscape, the patient autonomy movement and increasing cultural pluralism have also played important roles in the historical evolution of this issue, but have received far less attention. Over the past several decades, physicians have witnessed a shift from widespread, unquestioned, doctor-knows-best medical paternalism to the opposite extreme of patient autonomy. For most of the history of medicine, patients followed the advice of their physicians without a dialogue regarding alternatives, risks and benefits, or the patient's goals. The doctor was the expert and the patient bore the dependent, vulnerable sick role.[11] Physicians introduced only those treatment options they deemed appropriate, which, of course, left very little room for conflicts, especially in an era when patients had almost no access to medical information on their own. Contrast that with today's patient, who is nearly bombarded with easily accessible, if not always entirely accurate, information on the Internet, commercials, and television medical dramas. For years, physician beneficence was treated as the supreme guiding principle of medical ethics, and that was not challenged until the 1960s, when patient autonomy began to be asserted as a principle of equal or greater weight.[12]

As the locus of decision making has shifted from the physician to the patient, the patient's right to refuse care, which was the original spark behind the patient autonomy movement, has developed into a demand to receive the care of one's choice and also to dictate the precise details of how that care is delivered.[13] In many ways, we have moved from an era of physician paternalism to its polar opposite, in which physicians are viewed simply as the patient's agent and are expected to concede to patient demands without objection. For example, if parents want Ritalin for their child, they expect to get it; the physician's explanation that the child exhibits normal activity levels and does not suffer from ADHD may fall on deaf ears, for the parents have already made up their minds. If this doctor will not prescribe the desired drug, they can certainly find one who will. Similarly, if a patient has witnessed several friends taking a prescription drug with success, has viewed advertisements touting the drug's benefits, and has even researched the matter on the web, that patient will likely be taken aback by the doctor's position that the drug is just not right for him. This result may be received no more favorably than a store clerk refusing to sell a customer a suit because, in her opinion, it does not fit well. While the phenomenon of well-informed patients certainly has its upsides, many physicians have begun to complain that they are forced to waste valuable time convincing patients that what they want is not what they need.[14]

At this extreme, the physician seems forced to abandon her own moral agency in order to fill a new role as the patient's "technical accomplice," now responsible for simply using her technical skill to accomplish anything desired by the patient, within legal boundaries, regardless of the physician's moral qualms.[15] Clearly, the patient rights movement has created a serious, and not entirely beneficial, challenge to the autonomy of physicians. This at least partially explains why the physician's personal and professional beliefs about what is right are subject to greater attack today than ever before. Thus, protective conscience clause legislation can be viewed not only as a reaction to *Roe*, but also as a reaction to ever-expanding patient autonomy, a response through which the medical profession has attempted to reclaim some of its lost power.

The current debate surrounding these laws can also be understood as the latest manifestation of the struggles between religion and secularization in modern America.[16] As pluralism and cultural diversity have increased alongside the introduction of technological innovations and new bioethical questions that often demand members of a pluralistic society to reach some level of consensus as to how to move forward, the potential for conflict has become increasingly apparent.[17] Moral pluralism is a good thing, since, as Rawls noted, it indicates a genuinely free society in which individuals are permitted to pursue their own conceptions of the good.[18] Nonetheless, in recent years, American society has attempted to avoid the seemingly inevitable conflict resulting from the breadth of cultures and beliefs coexisting in this nation largely by asking everyone to restrict their beliefs to their personal lives.

This sort of secularization is an apparent attempt to demonstrate tolerance for disparate perspectives, but in reality it is disingenuous, for tolerance is comfortable and painless when one is never forced to confront opposing views. Further, the segmentation of one's personality demanded by secularization may be utterly impossible and can lead to widespread dissatisfaction, particularly considering the fact that it appears to "tell even devout people to treat religion as a once-a-week, private activity—in tension with the view that religion affords a complete way of life."[19] This has resulted in a backlash against secularization and the increasing presence of "private" views in the "public" sphere visible today. For example, in 2000, 70 percent of Americans wanted religion's influence in this country to grow,[20] and further denunciation of secularization is evidenced by the political organization of the religious "right" and the increasing acceptability (or at least expectation) of religious discourse in politics.[21]

Combining the effects of the patient autonomy movement and pressure toward secularization, religious physicians, and those with deeply held moral beliefs, are beginning to feel marginalized. They see themselves as instructed to ignore the very beliefs that may have played a vital role in their choice to enter the noble healing profession, expected instead to concede all control to patients, whose beliefs are to be almost unconditionally respected. For some physicians, this one-sidedness seems far too much to ask.

Many claim that they are truly not attempting to impose their beliefs on anyone and are not even purporting to know better than the patient what is in the patient's best interests. Instead, they are simply seeking protection of their *own* moral integrity in the face of increasingly questionable technological advances and patient demands.[22] Others can do it—and perhaps others should do it—but they themselves cannot. In this regard, the expression of physician conscience need not be about power or necessarily about forcing one's personal views on others.

Of course, not everyone is willing to accept these physicians as benign refusers, suggesting instead that their refusals, at least with regard to abortion and contraceptives, are based on a belief that women seeking these services are "promiscuous at best, and potential murderers at worst," which in turn is rooted in deeper beliefs about the role of women in society and the family.[23] For these skeptics, refusal at its core is all about the control of others. This may in fact be an accurate portrayal of the motives of some physicians involved in the current debate, physicians who see their conscientious beliefs as expressing the revealed truth of what is universally good for all based on divinely established categories of right and wrong. Not only will they not provide the service in question, but, these doctors argue, neither should anyone else.

Regardless of whether refusers view their conscience as a reflection of nothing more than their own idiosyncratic understanding of morality or as a reflection of God's absolute standards applicable to all,[24] the impact of their behavior on the patients dependent upon them for care may nonetheless have the same effect, and consequences are often what matter most. Avoiding these negative consequences of conscientious refusal will be the essential task for any acceptable compromise solution.

What Are Conscience Clauses Doing for Physicians?

While pharmacists and institutional refusers, such as hospitals and insurers, have received a great deal of attention in the current debate, it may seem odd that there has not been more significant discussion in the media, courts, and legislatures about physician refusals in particular. Surely doctors are not immune to these conflicts, but their stories are chronicled with far less frequency, though there is no clear explanation as to why that is the case. There is some notion that patients who are denied treatment rarely complain, perhaps because the situation feels too personal and humiliating to make public, and that when patients do make noise, things are often resolved quietly.[25] Of course, these theories are powerless to explain why pharmacists' refusals have garnered so much attention, particularly since those stories deal with similarly personal sexual issues.

Academic discussion of physician refusers has been somewhat more robust, but still not to the extent one might expect. Medical ethicists and philosophers have addressed the issue of conscience as applied to physicians, but even they have tended to focus more on other instantiations of the debate, and none has offered a convincing and feasible solution to conflicts of conscience that would protect both patients and their doctors. Further, the legal literature suffers from a surprising paucity of commentary evaluating the propriety of the pervasive laws that protect physicians from nearly any imaginable consequence of refusing to comply with patient requests on moral grounds; discussion of refusals by pharmacists, hospitals, and insurers is far more widespread.

On the litigation front, however, there has been some interesting case law directly involving physician refusals. For example, an ongoing case in California, which will be discussed in greater detail in chapter 6, involves the denial of reproductive services to an unmarried lesbian woman. The state supreme court will assess whether the physicians' behavior, allegedly based on their religious beliefs, was constitutionally protected, but the state's conscience clause is not directly implicated, since it covers only abortion services.[26]

In 2004, a Pennsylvania court held a physician liable for medical malpractice not as a result of his refusal to perform an abortion, which was protected by statute, but rather due to his failure to provide full informed consent by not disclosing the dangerous situation facing the mother so that she might have chosen to seek the procedure elsewhere.[27] A much earlier case in Washington was similarly willing to allow a physician's refusal of services on moral or religious grounds, so long as the physician provided patients with all material information.[28] New Jersey courts have allowed physicians to refuse to participate in a course of treatment selected by the patient that the physician regards as inappropriate or disagreeable so long as the physician continues to provide basic care, reflecting the standard understanding of how a doctor can go about terminating a relationship with a patient.[29] Doctors have also been permitted to refuse bloodless surgery to Jehovah's Witnesses as a result of claimed medical standards without being held liable,[30] and have even been allowed to refuse care as a result of their personal economic beliefs, just as long as patients were notified of this policy in advance.[31]

Despite this breadth of cases, however, there has been a dearth of reported opinions specifically dealing with challenges to, or even addressing, conscience clause statutes as they apply to physicians. Why the silence? One explanation for this apparent anomaly is the simple fact that physicians are unlikely to enter a field where they will predictably face personally objectionable situations—a selectivity that is not available to generalists like pharmacists, insurers, and hospitals—thus eliminating opportunities for conflict and the pursuant litigation or discussion. Of course, as technology moves forward, selectivity in one's choice of medical fields will likely be less effective in avoiding conflicts, particularly as embryonic stem cell therapies develop, impacting everyone from oncologists to endocrinologists to gerontologists.

Another part of the explanation involves important factors surrounding the creation of the doctor-patient relationship. Existing case law conveys the well-established rule that initiation of this relationship is entirely voluntary for both parties; the express or implied consent of the physician is required in the form of some affirmative action toward treating the patient.[32] Thus, according to legal analysis, as well as statements of professional ethics adopted by the American Medical Association,[33] physicians are free to refuse to accept a prospective patient for any reason not prohibited by law or contract, and could therefore refuse to take on a patient with whom the physician foresees a conflict of conscience arising. Because physicians have no duty of care to nonpatients, prospective patients who are refused services based on a physician's moral beliefs have no basis on which to bring a lawsuit, offering physicians a level of protection completely outside the realm of conscience clause legislation. This effective preemption of claims may be responsible for preventing widespread discussion of conscientious refusal in medicine by the courts and the broader public, despite the fact that allowing physicians such vast discretion in patient selection fails to address potential access problems that could become quite problematic as grounds for conscientious refusal expand in line with technological developments.

Once a doctor-patient relationship has been established, patients are still unlikely to attain a successful remedy against a physician refusing to provide a service based on his or her personal moral objections. As a threshold matter, the patient may not even be aware that a refusal is occurring, particularly in light of evidence that a significant number of physicians consider it acceptable to withhold information about medical services they find objectionable.[34] That issue aside, establishing that a physician had a duty to initiate, or even to complete, a service may be quite difficult, since a physician can terminate an existing relationship at any time so long as he provides the patient with sufficient notice of his intention to do so.[35] Further, while patients may experience inconvenience as a result of a physician's conscientious refusal, they may not be so inconvenienced as to render the pursuit of a lawsuit or disciplinary

complaint worthwhile. Additionally, there is always the possibility that
the patient disagrees with the physician's behavior, but nevertheless
respects the physician's decision based on a recognition that the requested service is morally controversial and not everyone will be comfortable providing it.

More important, depending on the scope of a state's conscience clause,
physicians' refusals may be protected with regard to the services they are
most likely to refuse to provide, essentially eliminating any duty that
could be used to form a malpractice case against them. The Illinois statute, for example, provides that "no physician or health-care personnel
shall be civilly or criminally liable to any person, estate, public or private
entity or public official by reason of his or her refusal to perform, assist,
counsel, suggest, recommend, refer or participate in any way in any particular form of health-care service which is contrary to the conscience of
such physician or health-care personnel."[36] As we will see shortly, these
statutes are hardly amenable to legal attack, so affected patients have little incentive to launch a judicial challenge—they are unlikely to succeed
in an attempt to invalidate the conscience clause, so they will also fail in
any attempt to hold the protected refuser liable.

On the employment front, conscience clauses offer significant protection that might not otherwise exist. As we saw above, the federal Church
Amendment protects individuals employed by institutions receiving federal funding from employment discrimination, discrimination regarding
staff privileges, and discrimination in admission to training and study
programs, while several state statutes offer an even broader shield.[37] In
fact, aside from the end-of-life cases involving refusal to withhold or
withdraw life-sustaining medical care, which are often suits against institutions,[38] most cases involving the conscientious refusals of individual
providers have been employment law cases.

In these suits, the health-care worker sues his employer for firing or
demoting him after he has refused to provide a given service for moral
or religious reasons. The matter is resolved through reliance on either
conscience clause statutes or Title VII, a part of the landmark Civil
Rights Act of 1964 that prohibits discrimination on a variety of bases
including religion.[39] This may at least partially explain why doctors
have not traditionally been involved in lawsuits on these grounds and

why nurses and pharmacists deal with this problem far more frequently: doctors are quite often self-employed and are more often explicitly protected by state law from negative employment consequences or loss of privileges related to their refusals. In the end, the statutes create a cause of action for physicians (and possibly for other health-care workers), but eliminate one for patients, if they ever had one at all.

This combination of factors offers a less-than-satisfying account of why public commentary specifically regarding physician refusals has not been more prevalent, since it seems that we should be talking more about whether conscience clauses applicable to physicians are appropriate or are in need of revision. But it does at least help to explain the state of existing case law. At first glance, the fact that doctors do not appear to be getting sued, fired, or disciplined for refusing to perform services on moral grounds might appear to indicate either that such refusals are not in fact a problem or that physicians do not actually need explicit protection of their conscience because it is available from some other source. However, that is emphatically not the case. While doctors may be able to avoid liability through careful rejection of potential patients and may glean some limited employment protection from Title VII, existing conscience clauses are doing something for physicians. Their most important role, however, may be concealed by the status quo.

The primary value of conscience clauses for physicians becomes visible upon consideration of the limitations on conscience that state licensing boards could feasibly impose in the absence of these statutes, which frequently prohibit licensing boards from discriminating against conscientious refusers. Under their police powers, states have broad authority to regulate the conduct of the medical profession, including the elements of training and capacity required for permission to engage in the practice of medicine. They have largely delegated this power to state medical boards, which determine the qualifications required to practice medicine in the state, including medical school graduation, postgraduate training, passage of various examinations, and the like.[40] Further, state medical practice acts authorize licensing boards to take disciplinary action against physicians engaging in unprofessional or dishonorable conduct, which the Federation of State Medical Boards suggests should be defined to include at least

(4) conduct likely to deceive, defraud or harm the public; (5) disruptive behavior and/or interaction with physicians, hospital personnel, patients, family members or others that interferes with patient care or could reasonably be expected to adversely impact the quality of care rendered to a patient; ... (9) negligence in the practice of medicine as determined by the Board; ... (12) practice or other behavior that demonstrates an incapacity or incompetence to practice medicine; ... (35) violation of any provision(s) of the medical practice act or the rules and regulations of the Board or of an action, stipulation or agreement of the Board; (36) engaging in conduct calculated to or having the effect of bringing the medical profession into disrepute, including but not limited to, violation of any provision of a national code of ethics acknowledged by the Board.[41]

Any one of these provisions could be read broadly to require a physician to provide even those medical services to which he or she is morally opposed. Physician refusal might conceivably be construed as unprofessional, likely to harm the public, disruptive behavior that interferes with patient care, negligence, incompetence to practice medicine, or conduct that could bring the profession into disrepute.

Without conscience clauses, licensing boards could take an even more direct route to limiting physician conscience. For example, in 1987 the New Jersey Board of Medical Examiners prohibited licensees from categorically refusing to treat a patient who was HIV+ or had AIDS when the physician possessed the skill and expertise to treat the condition presented.[42] While this was a licensing requirement based on certain types of patients, absent any existing statutory protection against professional discipline resulting from conscientious refusal, it is possible that boards, prompted by patient access concerns, might demand that physicians provide certain services regardless of their moral objections as a condition for obtaining or retaining their license to practice. In fact, Nora O'Callaghan describes a hypothetical state law requiring all physicians licensed by the state to provide abortions without any basis for exemption, even on religious grounds, after a finding by the state that many of its counties lacked abortion facilities and that women were thus unable to avail themselves of their fundamental reproductive rights.[43]

The question of whether such a statutory or regulatory obligation on individual professionals to provide abortion services (or any other medical care) upon request would survive a First Amendment challenge is a complex one, and it is largely fact based. The Supreme Court has imposed various levels of scrutiny when reviewing claims that particular laws violate the First Amendment. In the free exercise context, it has upheld laws of general applicability that are neutral with regard to religion so long as the law has a rational basis. In *Employment Division v. Smith*, for example, the Court relied on precisely these grounds to reject a claim challenging a denial of unemployment benefits to plaintiffs who had been dismissed from their jobs as a result of religiously inspired peyote use.[44] In other First Amendment contexts, however, such as those involving laws apparently targeted at religion (rather than simply creating an incidental burden) and challenges based on the freedom of association or speech, the Court has engaged in a more stringent analysis, asking whether the law's restriction of a fundamental right is necessary to the accomplishment of a compelling government interest.[45]

Because of the intricacy involved in First Amendment analysis, we cannot definitively conclude that O'Callaghan's hypothetical statute could withstand constitutional scrutiny,[46] particularly since it involves an affirmative requirement to do something rather than just a restriction on behavior that an individual might otherwise wish to undertake. What we can say, however, is that what she describes is a neutral law of general applicability that has the legitimate goal of improving patient care— the abortion service mandate would apply to all physicians and is not intended to target any particular religious group or practice, although it may have that secondary effect. We can also say that the government has a compelling interest in ensuring widespread access to reproductive medical services, although in light of the proposal explored in this book, we will see that it may become more difficult to argue that the goal of patient access could not be achieved by less restrictive means.[47]

The fact-intensive nature of this area of law is demonstrated by a 2007 decision from the U.S. District Court for the Western District of Washington involving state pharmacy regulations, which will be discussed in more detail in chapter 4. For now, suffice it to say that these regulations do not explicitly bar pharmacists from refusing to fulfill legal prescriptions on grounds of conscience but preclude pharmacies from doing so. In the context of granting a preliminary injunction against enforcement, the court held that the regulations were neither neutral nor generally applicable, despite the facial satisfaction of both criteria.[48]

Although the regulations apply to all pharmacists and pharmacies in the state, and to all types of prescriptions, the fact that they were explicitly prompted by religious refusals to dispensing "Plan B" emergency contraception led the court to conclude, somewhat questionably, that policymakers acted with the direct intent to burden free exercise rights. Additionally, although the regulations were overtly an attempt to improve access to a wide variety of prescriptions, a public welfare rationale that the court acknowledged could normally suffice to overcome the increased constitutional scrutiny leveled against laws intentionally and uniquely burdening religion, the court was not convinced that this was the true motivation here. Not only are there regulatory exceptions that allow continued inhibition of patient access on secular grounds, but the court also stressed that it had seen no evidence that access to care had truly been eliminated by the religious refusals of pharmacists. Thus, the court determined that on these facts, enforcement of the regulations would likely violate the First Amendment.

While this preliminary opinion of a lower court should not be given much weight, the case does at least indicate that the language, breadth and intent of an imposition on licensed health-care professionals to provide certain services can make all the difference. However, it certainly does not go so far as to suggest that any such service requirement would be inherently constitutionally impermissible.[49] In fact, the Court of Appeals of New York has upheld a law somewhat similar to that described by O'Callaghan, albeit one regulating institutions rather than individuals, that requires all employers who choose to provide insurance coverage for prescription drugs to also provide coverage for contraceptives, unless the employer's organizational purpose is to spread religious beliefs, it primarily employs persons sharing those beliefs, and it qualifies as a nonprofit organization. This second condition prevents many faith-based organizations from falling within the statutory exemption, but the court found that the law is nonetheless permissible under both the federal and state constitutions since it is facially neutral and its primary (and genuine) purpose is to improve women's health rather than to constrain religious activity.[50] The court also relied heavily on the fact that the plaintiffs were not required to provide prescription drug coverage at all, so they were not actually required to violate their religious beliefs.[51]

In this vein, it is essential to recognize that the courts have generally been less sympathetic to religious freedom claims made by those engaged in voluntary commercial activity, such as the practice of medicine, as opposed to directly religious activity, since the religious person could have theoretically made different professional choices that would not burden his or her beliefs at all.[52] For example, while a person could not be excluded from a state-licensed profession simply because he or she belongs to a particular religious group, the Supreme Court refused to invalidate a requirement that those wishing to practice law in a particular state take an oath to support the state's constitution, despite the fact that the oath included an indication of willingness to perform military service.[53] If that was not a violation of federal constitutional protections of religious freedom, there is significant reason to believe that an analogous demand of willingness to perform certain medical services could also be a permissible imposition on physicians.

While this sort of care mandate would likely face political difficulties in state legislatures, the growing public dissatisfaction over refusals by health-care professionals to satisfy reasonable patient requests suggests that the adoption of such a requirement is not entirely unlikely. After all, physicians' general freedom to choose their patients and decide when to terminate their relationships is already constrained by laws and regulations prohibiting discrimination and otherwise protecting patients. These restrictions could likely be expanded to add conscientious objection to the list of unacceptable physician behaviors.

Beyond this apparent *permission* to avoid conscience clause protection, however, some argue that it *must* be avoided. These commentators claim that the coupling of conscience clauses with state licensure of health-care professionals should be regarded as impermissible state action in the denial of the constitutional right to be free from state interference in reproductive choices.[54] However, similar arguments have been largely rejected by the courts in other contexts. For example, the Supreme Court has held that heavy regulation of a private utility company and provision of a partial monopoly by the state were insufficient to render the utility a state actor,[55] and also that regulatory oversight by a state liquor board did not render the discriminatory actions of a private club state action.[56] Because conscience clauses simply allow professionals

to choose for themselves, but do not actually prohibit the choice to provide legal medical services, they are distinguishable from impermissible
affirmative actions by the state to restrict reproductive autonomy or
other protected freedoms. Notably, over the several decades of their existence, no state or federal conscience clause has been struck down as
unconstitutional.

Whereas extant conscience clauses protect physicians from patients
and employers, and most especially from licensing boards that might
otherwise be able to exclude refusers from the practice of medicine entirely, we have now seen that these laws can be changed. Conscience
clauses appear to be neither constitutionally mandated,[57] nor constitutionally prohibited, leaving state legislatures with ample room to strike
a balance between the interests of both physicians and patients. Unfortunately, current conscience clause policy strikes the wrong balance—in
fact, it offers hardly any balance at all, allowing physicians to refuse in
too many situations without exception and without concern for the
patient's ability to access medical services. Of course, legislatures should
not move to the other extreme, imposing the sorts of blanket requirements on physicians described above, but instead should seek an appropriate middle ground. Thus, the crux of this book is about finding that
middle ground, and the policy it proposes is unlikely to run into constitutional resistance.

When *should* physicians have a duty to satisfy patient requests, when
should they be able to refuse, and how *should* access concerns most appropriately be addressed? These are the crucial questions, but to answer
them, we must first dispel some important misconceptions.

Where the Conscience Clause Debate Has Gone Astray

The battle lines have been clearly drawn, but unfortunately, responses to
the conscience clause controversy tend to suffer from several fundamental problems. First, much of the existing discussion has inappropriately
narrowed the issue to a focus on religious beliefs and associated objections. It often fails to recognize other important types of conscience,
such as secular moral beliefs or, more important, understandings of
professional ethics that also have a crucial role to play in the clinical

encounter between doctor and patient—and have the same, if not greater, potential to create access problems. A much broader conception of conscientious refusal is needed, one that will include refusals grounded in values that are widely held within the profession and have even been accepted as clinical standards, but that are not based exclusively on the profession's technical and scientific expertise.

For example, consider the surgeon who refuses to accept a patient unless that patient first agrees to undergo a screening procedure that carries some significant risks of its own in order to rule out greater risk factors that could severely complicate the surgery. That doctor is only looking out for the patient's safety,[58] but what he or she is effectively saying is "regardless of the risks you as a patient are willing to accept, I cannot in good conscience impose those risks on you." This behavior is likely in accord with professional standards, and those standards are what drive the physician's refusal, rather than some religious objection to the surgery itself. However, that other doctors would also refuse does not diminish the fact that the refusal is nonetheless based on ethical beliefs about the level of acceptable risk that the patient may not share, rather than on some objective, indisputable determination as to which risks are too great.[59] Therefore, this is a conscientious objection, albeit one made at a professional level and labeled as an issue of professional integrity or the internal morality of medicine. The fact that it is made by a group rather than by an individual does not necessarily suffice to render it acceptable, and as we will see, may make it even less so. But this is precisely the sort of refusal that would evade consideration under the narrow, stricter understanding of conscience that has prevailed over most of the existing debate.

In other ways, the matter of conscience has been treated too expansively. As Martha Swartz notes, more than one-third of U.S. conscience clauses fail to state what constitutes acceptable grounds for refusal, and some are drafted so broadly or ambiguously that they could "equally protect the right of a health care professional to refuse to participate in a medical treatment because the procedure was scheduled too early in the morning or because the procedure was controversial."[60] These grounds for refusal are likely already protected by the laws governing the doctor-patient relationship discussed previously and may in fact be important, given that the ability to select one's patients and to define the scope of one's practice are both crucial freedoms that contribute to the establishment of a personally fulfilling career. However, Swartz is right to point out that they are not true expressions of conscience and should be dealt with separately—but how exactly can we differentiate conscientious refusals?

The sorts of refusals that appropriately fall under this heading may in fact be quite broad, and conscience itself is a slippery concept that has been defined as the ethical tug toward doing the right thing that becomes a central, dominating feature of one's motivation and self-identity.[61] It is "the interior, quintessentially human voice that speaks to us of goodness and duty, the voice we must obey if we are to keep our integrity."[62] Descriptions of conscience often boil down to fuzzy claims that a person could not be convinced to do the thing in question for any price and to notions that if a person did some act, he just could not live with himself, look himself in the mirror, or fall asleep at night. However, there may be different intensities of true conscience that these descriptions do not seem to recognize.

Note that some commentators attempt to draw a further distinction between objections rooted in general moral grounds and objections truly based on conscience. For example, Kent Greenawalt describes a nurse who believes that plastic surgery is inappropriate because it is materialist and superficial, encourages cultural denial of aging, and wastes valuable resources. He suggests that she may have "moral reasons not to help in such operations, but the reasons do not amount to a conscientious objection, and they might not render her assistance an act against conscience."[63]

While this reasoning appears to focus on how deeply held the personal beliefs in question are, perhaps in an effort to address the intensity issue raised above, the line Greenawalt attempts to draw is not self-evident and, in fact, seems to get things wrong. Acting contrary to these beliefs *would* damage the nurse's integrity, though maybe less significantly than participating in some action that she considers to be an even greater wrong. Perhaps more importantly, even if not a crucial, inviolable part of her self-identity, this sort of objection contributes to the critical moral debate regarding the proper ends of medicine, the value of which is discussed in detail in chapter 3. Thus, for our purposes, the phrase "conscientious refusal" ought to be defined broadly to include nearly every *normative* ground for objection to a medical service, even if philosophical definitions of conscience have traditionally been more restricted. However, we will see that not every type of conscientious refusal should be protected simply by virtue of being defined as such.

In contrast, refusals on *non-normative* grounds, such as personal convenience, accommodation of popular opinion merely to avoid controversy,[64] economic considerations regarding which services are reimbursed at the highest rates, or mere aesthetic distaste,[65] do not satisfy either of our criteria for conscientious refusal. They would not involve regret or self-loathing if the physician was forced to act, nor do they necessarily contribute to the important social debate about whether medicine is veering into ethically troublesome territory. Thus, two irksome cases can be excluded from our discussion entirely: the distasteful or inconvenient service and the difficult patient.

If a physician who finds abortions to be terribly boring but not morally objectionable has to perform one,[66] he may be disappointed or even frustrated, but he would not feel guilty about what he had done and would certainly not lose any sleep over it. Potentially more telling: if that physician were offered enough money, he might be tempted to make abortion procedures the entirety of his medical practice. Further, he is not making any socially valuable statement about whether abortions are good or bad, right or wrong, admirable or deplorable—instead, his conscience is morally neutral. Similarly, an obstetrician who just hates to get up in the middle of the night might opt only to perform cesareans or induced births that he could schedule according to his plans, but could assist in an unscheduled birth without compromising his moral integrity.

Likewise, a doctor faced with a particularly belligerent patient who misses appointments, chastises staff when he does show up, and generally detracts from the quality of care available to other patients, could choose to end the relationship because it is just too challenging. For example, in *Payton v. Weaver*, a California court held that a physician had no obligation to continue providing dialysis treatment to a patient who was disruptive and uncooperative in a way that affected other patients and staff.[67] Nevertheless, continuing to provide such a patient with care would not necessarily be *wrong* from the doctor's perspective, and thus would not violate his conscience. In fact, refusing to treat such a difficult patient may be more ethically troublesome for the physician, particularly if he worries that the patient may not be able to access needed care from anyone else.[68]

These nonmoral reasons are important and may in fact explain far more refusals of care (and potential access problems) than what can properly be understood as conscientious refusal, even on the broadest account. Whether or not they are valid and appropriate is an open question, but they are distinguishable from conscientious refusals and thus present unique questions that will not be addressed here.

Aside from these issues of scope, another significant problem with much of the existing commentary is that despite the fact that conscience clauses have expanded their reach in the decades since *Roe*, the debate has unfortunately never shed its abortion roots. In fact, several observers have labeled the issue a mere proxy for the quintessential culture war that is the long standing abortion controversy, noting that several of the main arguments have simply been recycled and are thus similarly destined for stalemate.[69] However, it is simply not the case that opponents of abortion will always be proponents of protecting conscientious refusal or that those who are pro-choice will be resistant to allowing doctors to avoid the provision of certain services. For example, a physician could be wholly opposed to withdrawing life-sustaining treatment from a patient in a persistent vegetative state, since he may feel compelled to save lives whenever possible, rather than to permit death. Nevertheless, the same physician may have no moral objection to abortion, perhaps because he does not believe that a fetus has personhood rights such that the procedure is still in line with his understanding of his professional obligations. A doctor could also foreseeably refuse to provide hormonal stimulation to an infertile woman who will not agree to selective reduction of high multiple pregnancies should that occur—that doctor would be refusing not to abort. This clearly demonstrates the difficulty with the proxy approach, for the arguments surrounding conscientious refusal and abortion will not always run parallel to one another.

The abortion focus is inaccurate and misleading in other ways as well, at least in part because it underestimates the nature of the problem. It may be the case that the archetypal image conjured in the minds of most people on hearing the phrase "conscientious objection" applied to the medical context is the physician (valiant or derelict, depending on one's view) who refuses to perform an abortion. However, directing attention exclusively to that type of refusal will obscure a variety of other important breeding grounds for conflict. What about the physician who objects to assisted suicide or euthanasia should those acts be widely legalized or who objects right now to the withholding or withdrawal of life-sustaining care,[70] who refuses to satisfy a patient's request for care that he believes would prove futile, or who will not operate on a patient who rejects blood transfusions because without them the surgery would be too dangerous? What of the doctor who will not supervise the lethal injection of prisoners or who refuses to monitor political detainees to ensure that they can continue to be tortured without dying?

These examples are only the tip of the iceberg, for technologies derived from embryonic stem cell research, procedures involving genetic manipulation, assisted reproduction, enhancement technologies, and other medical advances will greatly expand the breadth of procedures that physicians find morally objectionable. As Martha Swartz has noted, the combination of new technologies entering the medical scene and the increasing diversity of health-care providers has the potential to create a perfect storm: "It is likely that at least *some* health-care providers may object to the application of *some* technology to *some* patient on the basis of *some* religious or moral belief."[71] The bottom line is that more doctors and more patients are likely to find themselves affected in the relatively near future.

Thus, the stakes are high, and limiting the debate to tired abortion rhetoric could be quite dangerous if it prevents meaningful discussion of the broader propriety of physician refusals. Without a doubt, abortion is intricately involved in this issue, but it is essential that the conscience clause debate extract itself from that tangle. Otherwise, true compromise will remain impossible as both ideological extremes continue to shout past one another; indeed, that is precisely what we have seen so far. Instead, the debate can be more appropriately recast as an

issue of professionalism focused on clearly defining professional obligations, nothing more and nothing less.

While the myopic focus on abortion is a serious flaw hindering progress in this area, by far the biggest problem confronting the current
debate over conscientious refusal in medicine is its uncritical, almost sen-
sationalist, framing of the issue as one of competing rights.[72] Patient
advocates talk about their right to access desired, legal medical services,
while those supporting the cause of some health-care professionals focus
on their right to exercise personal conscience even while wearing their
professional hats.[73] While it is unclear whether these arguments are referring to legal rights, as opposed to moral rights that could be far
broader in scope, an unfortunate consequence of this competing-rights
language is that it makes it quite difficult to reach any resolution.

When the debate is focused solely around rights, it is difficult to avoid
the problem of rights as trumps—both sides of the debate claim important rights that defeat those of the other simply by virtue of the fact that
they are rights.[74] As Jeremy Waldron and Thomas Hobbes have elo-
quently noted, "for people to demand that we treat *their* theory of rights
as the one that is to prevail is 'as intolerable in the society of men, as it is
in play after trump is turned, to use for trump on every occasion, that
suit whereof they have most in their hand.' "[75] Of course, this leaves no
room for reasoned argument, explaining the all-or-nothing nature of
many proposed solutions to the conscientious refusal dilemma.

More problematic, if patient advocates are using the term in the legal
sense, is that American patients actually have very few rights that they
can demand from health-care providers, rendering the rights-based argument insufficient to vindicate the sort of broad patient claims that have
characterized the debate thus far. There seems to be some misconception
that simply because something is legal (i.e., not prohibited), we must
have a right to it. However, this sort of analysis proves far too much
and is based on a failure to differentiate between positive and negative
rights. If we understand positive rights to impose correlating duties or
responsibilities on others to ensure that those rights can be exercised,
then claiming a positive right to all that is legal cannot work. For example, it is legal for you to purchase your favorite orange juice, but if
the company that manufactures that juice goes out of business or your

local grocer no longer sells the product, you have not been denied any right, for no one owes you any obligation to provide that particular orange juice. Legality on its own simply does not mandate universal availability.[76]

If refusing physicians are in fact violating patient rights, something else must be at play, though, as we will see, it cannot be found in any legal source. First, patients have no clear constitutional rights to any affirmative care,[77] and certainly no right to affirmative care from any particular health-care provider.[78] Whether normatively appropriate or not, we saw previously that a woman has absolutely no positive right to abortion, but instead has only a negative right against unduly burdensome state interference in the choice to seek an abortion. Private behavior is not implicated at all,[79] and courts have repeatedly held that public institutions are not required to facilitate abortion in any way, including by offering or paying for it, and can even try to persuade women to choose other options.[80] Therefore, a physician's refusal to perform an abortion based on his or her own moral objections to the procedure violates no constitutional rights of the patient. Similarly, *Griswold* establishes no positive right to contraceptives, but rather provides a more restricted negative right against state laws prohibiting their use or sale.[81] Finally, while patients do have constitutional rights to refuse medical treatment,[82] they have no positive right to demand that any individual physician be the one to terminate treatment that has already been initiated.[83] As demonstrated by these examples, virtually all constitutional rights, especially when it comes to medicine, are "freedoms from" rather than "freedoms to." Thus, they cannot provide sufficient support for claims of patient rights against conscience clause protection for physicians.

Of course, the Constitution is not the only source of rights, and several statutes do provide patients with positive entitlements. However, these are relatively uncommon and certainly not pervasive enough to support a right against conscientious refusal. For example, the existence of Medicaid and Medicare benefit schemes creates positive entitlements to certain services for eligible individuals, but these entitlements are not physician-specific given that doctors are free to opt out of the programs. The federal Emergency Medical Treatment and Active Labor Act (EMTALA) offers another positive right, and requires hospitals that have

emergency departments and participate in the Medicare program to provide medical screening and stabilizing treatment to all emergency room
patients in a nondiscriminatory manner.[84] However, EMTALA does not
extend beyond hospitals, and only questionably applies to individual
physicians in covered emergency rooms.[85] Further, most states have no
duty-to-rescue, or Good Samaritan, statute,[86] so that physicians not covered by EMTALA have no legal obligation to aid a patient in need of
medical care absent a preexisting relationship between them.[87] Therefore, patients lack any legal right to emergency care from a particular
physician, although most codes of professional ethics encourage physicians to provide emergency treatment and many state conscience clauses
exempt emergency situations from the ambit of protected refusals.[88]

Patients do have a common-law right to demand that physicians obtain their informed consent before embarking on any particular course
of action,[89] and this right has been codified in many states.[90] Patients
also have a common-law right of referral when their physician knows,
or should know, that he lacks the requisite skill, knowledge, or facilities
to treat the patient's ailment properly,[91] although it is unclear that a conscientious refusal can be categorized in such a way. Finally, patients have
a common-law right to be treated in reasonable accord with the standard
of care and not to be abandoned by their physician once a doctor-patient
relationship has been established, but have no legal claim on any physician before that point.[92] Further, abandonment liability does not apply
to a patient not in current need of continuing medical care, and even if
the patient does require further medical attention, as alluded to earlier,
the objecting physician must only ensure that the patient has sufficient
notice of the physician's intention to terminate the relationship such
that the patient herself can procure other medical attention.[93]

The bottom line is that despite the arguments of conscience clause
opponents who allege that refusers are violating patient rights of access,
no such legal right exists. In fact, the lack of positive patient rights
to health-care services is one of the major criticisms of the American
health-care system. And because patients have so few legal rights,
health-care professionals have a great deal of open space in which to
exercise their own consciences even if they have no express legal right
to do so.

However, while nearly all states have some statutory protection for physician conscience, creating a true legal right in those protected contexts, the current analysis must ignore such statutes because the propriety of their very existence is one of the major issues analyzed in this book. More important, we saw previously that legislatures have constitutional latitude in this area and could likely eliminate conscience clause protection altogether. Further, while doctors do have a federal statutory right against employment discrimination on the basis of their religious beliefs rooted in Title VII,[94] that right is extremely constrained, requiring no more than de minimis accommodation,[95] and is of no use to secular refusers. Therefore, while physicians currently have at least more de facto protection than patients, it is on at least somewhat shaky ground. In the end, it becomes clear that the "practice of medicine is a privilege ... not a natural right of individuals."[96]

As should now be obvious, the legal rights starting point often used by both sides of the current debate does not provide solid, consistent, reliable protection of either party. It simply cannot do the work that is being asked of it, and a different paradigm for solving this dilemma is required. An alternative analysis of legal duties clearly will not get us very far, since if there are no legal rights, there are no legal duties. However, if we approach the problem in light of the *moral* obligations of professionals, as some commentators have done,[97] we will begin to understand what ought to be the case, rather than relying on what currently is, and can also avoid the problem of trumping. These moral duties will illustrate what moral rights both doctors and their patients can legitimately claim, which in turn will inform the analysis of what conscience clauses should really look like, and even whether they should exist at all. There are many conflicting and complementary understandings of a physician's professional responsibilities, but once we recognize that the obligations of the profession as a whole may not precisely correspond to the obligations of the profession's component members, we are on the road to an appropriate compromise.

2

Defining Medical Professionalism

Professionalism is an integral component of the conscience clause debate, which at its heart asks whether being a medical professional demands the willingness always to place others before oneself and whether personal or professional norms take precedence. However, professional obligations are an incredibly tricky concept to characterize, and nearly everyone, including doctors themselves, struggles to adequately and accurately explain the terrain. Unfortunately, they often can do little more than fall back on Justice Potter Stewart's definition of obscenity; they may not be able to define professionalism intelligibly, but they know it when they see it—or better yet, they know unprofessionalism.[1] It is unsurprising, then, that physicians, and their professional organizations, often disagree with one another as to which model of the doctor-patient relationship is most appropriate and with regard to what specific duties the physician owes the patient.

Clearly, a more precise definition is needed. Understanding the nature of the social contract in which the public grants the medical profession incredible control over the services and knowledge we need for our own happiness, health, and sometimes very survival, *and determining what the public is to receive in return*, will play a critical role in deciding whether conscientious refusers have truly shirked their duties, as they are so often accused of. In doing so, this endeavor will also lay the foundation for a feasible solution to the practical problems raised by physician refusals.

Discerning the Breadth of the Primacy Principle

According to one version of medical professionalism, what the public demands in return is that physicians

> subordinate their own interests to the interests of others, ... adhere to high ethical and moral standards, ... respond to societal needs, ... reflect[ing] a social contract with the communities served, ... evince core humanistic values, including honesty and integrity, caring and compassion, altruism and empathy, respect for others, and trustworthiness, ... exercise accountability for themselves and for their colleagues, ... demonstrate a continuing commitment to excellence, ... exhibit a commitment to scholarship and advancing their field, ... deal with high levels of complexity and uncertainty, ... [and] reflect upon their actions and decisions.[2]

Note that the conscientious refuser could comply with each of these elements of medical professionalism, except for *completely* subordinating her own interests to the interests of her patients—a strong version of the concept known as the primacy principle.[3] Importantly, however, complete subordination in each and every circumstance does not seem to be exactly what professionalism requires.

Though the American Medical Association is unsurprisingly inclined to protect its own members, its Code of Medical Ethics is often relied on as an important statement of professional obligations. The code provides that physicians have an ethical responsibility "to place patients' welfare above their own self-interest,"[4] but it simultaneously allows physicians wide discretion to refuse their services outside of emergency circumstances for nondiscriminatory reasons.[5] There are certainly arguments that allowing this sort of freedom is in the best interests both of patients, who would not be well served by a professional who did not want them, and physicians, who want to preserve some level of autonomy in their professional lives, especially when they can be held morally accountable for the actions in which they are involved on behalf of their patients. However, it would be inconsistent to claim that professionalism demands unyielding willingness to subordinate one's own interests to the interests of the patient, while concurrently allowing the refusal of one's professional services to a member of the public.[6]

Such inconsistency shows that the AMA has not accepted patient primacy as an absolute value, and neither have others. If patient interests were unconditionally supreme, it would be unproblematic that forty-seven million Americans are uninsured,[7] for example, since physicians would be obligated to provide free care regardless of their own financial needs. Doctors might also be chastised for (or possibly prohibited from) taking vacation days or selecting medical specialties based on how well they pay or what they find most interesting, since sacrificing their own preferences would be better for patients. Of course, most people do not intend to take things that far and the absurdities of these examples demonstrate that a weaker version of the primacy principle is in order. Physicians are not serfs, and if we treated them as such, we would be left with very few.

But how far does the primacy principle reasonably extend? Without defining the limits that must exist, reliance on the broad principle of altruism referred to in much of the existing commentary on professionalism does little to help us understand what sacrifices we can truly expect, and even require, of physicians wishing to remain in that professional role—and more specifically, whether we can demand that physicians sacrifice their interest in, if not their legal right to, freedom of conscience as reflected in their conduct. Reasonable minds disagree on the extent to which a physician can place her own welfare, or that of her family, above the needs of patients, and this disagreement about the professional responsibilities of physicians has only been exacerbated by changes to the profession itself. Notably, some of the same historical developments responsible for transforming the concept of what it means to be a physician mirror the factors responsible for sparking the conscience clause debate itself.

First, as discussed previously, the past several decades have seen a major transition in the traditional relationship between doctor and patient. Rather than a fatherlike physician giving directions to childlike patients, today we see a medical system in which patients and managed care organizations are competing with physicians for control. The emerging norm of patient autonomy threatens to erode the professional stature of medicine, reducing physicians to little more than "mere purveyors of medical technology."[8] The increasing development of a consumer-based medical system is also important, for if the relationship between doctor and patient is no more than a contract between equal parties, then the doctor has every right to refuse to become a party to a contract that would require violation of his conscientious beliefs.[9] The days in which doctors possessed enormous social privileges, were compensated in the currency of power and admiration, and had extraordinary control over patients are waning; that physicians' notions of extraordinary duty, in the form of house calls, charity care, and acceptance of personal risk, for example, have simultaneously deteriorated should therefore come as no surprise.[10] The quid pro quo is simply being rebalanced.

The technological developments that physicians may find themselves with moral objections to, especially at the beginning and end of life, have also had a dramatic impact on traditional understandings of what it means to be a good doctor and have raised serious questions about what a doctor's duties truly are.[11] What it means to heal a patient is no longer always apparent, nor is it obvious that a physician should do everything he or she can to preserve the patient's life, though these were the twin responsibilities of the medical profession for much of its history. For many years, technology was so limited that it was possible to attain general agreement on what was beneficial or harmful to the patient—certain outcomes and goals clearly stood out as best. But now medical technology is capable of so much more and this agreement has often dissipated.[12]

For example, is it better for the patient to be alive in a persistent vegetative state, or dead in accordance with his or her likely wishes? Is it better to abort a fetus discovered to have an anomaly that will prove fatal, or to allow the pregnancy to continue to birth, potentially allowing the child to suffer a more traumatic death as various mechanisms of technological support extend life without improving it? What the good physician should do with the technologies available is often as clear as mud, rendering a satisfactory exposition of patient interests and welfare, and concomitant professional obligations, even more troublesome.

The final challenge to conventional notions of medical professionalism is the fact that "the traditional doctor is not necessarily the modern-day doctor."[13] There has been increasing recognition that many physicians are no longer (if they ever were) willing to completely sacrifice themselves to their professional calling—requiring such sacrifice as part of what it means to be a physician is potentially outdated. As women have entered the workforce and as both sexes have recognized the value of work-life balance, the professions themselves have changed. Traditionally, many doctors tended to neglect other parts of their lives, such as their family and personal interests, and were taught that this was not only what was expected of them, but what would be demanded of them by their colleagues.

Today's physician, on the other hand, particularly the younger generation, is taking a stand against this lifestyle. The modern-day doctor "may be of either gender, work part-time and be the primary caregiver for his or her family. Or it may be a doctor who works part-time and runs a business as well, or does research, or maybe even one who has made a choice to work part-time and pursue hobbies and a balanced lifestyle. Such doctors now may regard themselves as having other important duties that compete with their duties to patients, such as those to their families, to their communities, to other jobs and maybe even to themselves."[14] In fact, Denise Dudzinski suggests that a physician choosing to reschedule a patient's appointment in order to go home to care for her own sick child behaves appropriately.[15]

Although it is still generally accepted that professionals should be willing to commit their working lives to the service of patients, the idea that physicians must dedicate their whole lives to their patients is increasingly viewed as unrealistic and not what we can or should expect. Once we recognize that the medical profession may still be a "calling," but that it lies somewhere between priesthood and just a job, sacrificing one's entire being to the demands of patients becomes clearly unnecessary. Most importantly, if physicians can refuse to see patients outside of emergency situations because those extra appointments would take away from family time, then it appears that physicians ought to be able to select patients in a way that avoids the necessity of compromising their moral lives or even their immortal souls for the profession.

A defining characteristic of the learned professions is that their members bear a special body of knowledge that entitles them to certain privileges, but that also saddles them with certain obligations. It may be the case that one of those obligations is "altruistic beneficence," which requires some level of preference for the intentions of others and some degree of effacement of personal interest.[16] However, the preceding discussion has made it clear that the appropriate balance between looking out for oneself and looking out for one's patients is often unclear and is very much in flux. What is clear is that there are different types of self-interest, some of which are legitimate for even professionals to pursue, such as duties owed to oneself to guard health and life, and duties owed to others, such as professionals' families. It is also clear that what physicians have a professional obligation to avoid is *illegitimate* self-interest, but the fundamental question that remains is whether concern for the preservation of one's personal moral integrity falls into that category.

Different models of medical professionalism have reached different answers to that question, as we will see below, but the right answer seems to be that it depends; a physician's refusal to provide care that violates his conscientious beliefs can only be judged legitimate or illegitimate based on the surrounding circumstances and the burdens such refusal will place on dependent patients. Importantly, these circumstances can be manipulated and burdens mitigated through imposition of the proper institutional obligations. There may be limits on when, how, and why a physician can engage in conscientious refusal, but demanding full renunciation of that interest is both unnecessary and overbearing.

The Consent Paradigm

In the context of exploring physician responsibilities during epidemics, Norman Daniels has used a model of professionalism based on a theory of consent that may be extended to other controversial areas of professional duty, such as those surrounding the permissibility of conscientious refusal, though Daniels himself has not applied his ideas in this way. Daniels uses the consent model as the just and appropriate mechanism for determining professional commitments, arguing that if physicians have a professional duty to undertake certain physical risks, it is only because they have agreed to do so. Becoming a physician is voluntary, and thus one's professional obligations as a physician must also be a voluntary undertaking. For Daniels, these duties do not stem from the mere concept of participation in a profession, nor from an understanding of the virtuous physician, but instead resemble something more akin to a specific contract.[17]

Despite his focus on voluntariness, however, Daniels objects to any extension of the notion of consent that would allow physicians to essentially define their own personal duties as professionals since that would eviscerate the very concept of professionalism, leaving patients without any standard responsibilities they can expect to have fulfilled by any given physician or member of a medical specialty.[18] In fact, he argues, as do others, that professional consent must be an all-or-nothing enterprise upon entry into the profession, at least with regard to a threshold below which professionals cannot dip, not one of selecting some personal obligations and rejecting others. Individual tailoring is an option only above some preset baseline such that all professionals have a duty to provide certain core services and to accept some standard level of risk;[19] "becoming a doctor means accepting *this* set of obligations," Daniels explains.[20] A physician can be appropriately reprimanded for failing to satisfy that set, but not for failing to undertake greater risks, which are supererogatory and deemed heroic.

Thus, the basket of services and risks to which physicians consent is determined not by physicians themselves individually or on a case-by-case basis, but rather by the socially negotiated ideal of the good physician. Individual doctors can instigate change through the proper social and professional channels, but cannot deviate from the services expected from their chosen specialty without broader change at the level of the entire professional community, or at least a significant portion thereof.[21] Notably, however, even commentators on professional ethics who generally accept the concept of social negotiation as opposed to individual deal making recognize exceptions to the rule: "Even in the individual case, the norms of the profession are not the ultimate determiners of right and wrong. If these norms are in conflict with one another or with other important moral considerations, or if they are severely defective in some way, then the professional must form his own conscience to decide how to act. Situations arise in which conscientious disobedience of a professional norm is what a person's moral judgment requires when all things about a situation are considered."[22]

As this discussion indicates, by voluntarily accepting their professional role, physicians have certainly agreed to relinquish *some* of their autonomy. However, we have still not settled the central question—exactly what do physicians consent to when they take on that professional role? That is precisely what we will now seek to determine.

When exploring the extent of physicians' obligations to treat patients despite the personal risk of infection, Daniels critically recognizes that the physician's choice to enter the profession did not in itself entail an agreement to face any and all levels of physical risk, no matter how high.[23] Instead, he and others agree that the "normal risk level of the working environment, the health-care worker's specialty, the likely harm and benefits of treatment, and the competing obligations deriving from the worker's multiple roles will all influence the duty of care,"[24] for each of these factors plays a part in determining the precise scope of the physician's consent. Understandings of physician obligations in the face of epidemics help to explain this point, and it is important to note that these have changed over the course of history.[25]

In ancient Greece and Rome, for example, physicians had no obligation to accept hopeless cases, and thus could avoid any duty of care if they had no therapeutic interventions to offer. By contrast, physicians who fled from the seventeenth-century European plagues were subject to social censure. Similarly, from 1847 to 1957, the AMA Code explicitly included an obligation on physicians faced with pestilence to continue treatment of the sick even if their own lives would be jeopardized. This responsibility was excised from the code fifty years ago, though by the late 1980s, the AMA (and some state licensing boards) felt the need to step in again by prohibiting the refusal of patients due to their HIV+ status.

Rather than raising the baseline, however, this particular provision was at root an attempt to hold physicians to what they could legitimately be seen as having already consented to. The true level of risk posed to physicians in HIV cases was on par with other risks to personal health that doctors had generally been willing to accept in the past, indicating that their reasons for refusal reflected nothing more than outright discrimination against the populations more heavily affected by the virus, or at the very least poorly informed views about modes of transmission.

Thus, even though a primary care physician practicing in rural Wisconsin may not have explicitly consented to treat AIDS patients when he entered the profession, potentially because the virus did not yet exist, he may nonetheless have implicitly agreed to undertake that professional obligation.[26] However, because greater personal risks may have been beyond the scope of consent, and thus beyond baseline professional responsibilities, it is unlikely that professional organizations would have imposed this responsibility had the risk of seroconversion in the line of duty been greater.

In fact, we can posit that any duty to treat patients suffering from SARS or avian flu, for example, could be appropriate only if physicians are provided with proper equipment and training sufficient to reduce the risk of personal infection to a level commensurate with that physicians can be seen as having already agreed to. Thus, in light of the difficulty of initial diagnosis, lack of training in the use of appropriate protective measures (and in some cases, their unavailability), and unpredictable variation in the ability of individual patients to spread the disease, physicians treating patients during the 2002–2003 SARS outbreak were frequently referred to as heroes, suggesting that they had gone above and beyond what was expected of them. What their duties truly were has been debated, but it is at least clear that today's physicians have no obligation to accept suicide missions.[27]

While these discussions of disease epidemics may appear at first glance to be immaterial for our purposes, they are enlightening because they acknowledge that there are some risks that physicians are unwilling to accept and that are not expected of them. Importantly, such hazards may include certain "moral risks." Moral risk is not something that has been addressed in the existing literature discussing either conscientious refusal in medicine or professional obligations in the face of personal harm, but its importance cannot be discounted. This is especially true given the stakes involved in demanding that a conscientious refuser provide the objectionable service, for in the eyes of some refusers, the consequences are beyond life and death—in fact, they may be no less than eternal paradise or interminable damnation. Given the impact immoral conduct can have on a person's spiritual destiny and personal integrity, as well as the emotional distress certain to accompany a decision to violate one's conscience, whether rooted in religious values or secular reasoning, moral risk may be as worthy of attention as the limited physical risks society asks physicians to assume. In some cases, they may be even more so.[28]

Physical risk is certainly more tangible, and its applicability is likely more homogeneous across all physicians—aside from the immunocompromised or the superimmune, physicians exposed to an infectious disease all face relatively similar personal risks, although some may be more willing to face those risks than others. Moral risks, on the other hand, are far more diverse in their applicability, since the very same service may pose a moral risk for one physician but not another. This distinction lends itself to the sort of argument that we will see later in this chapter, namely, that only those for whom the service poses no moral risk should be admitted into the profession. Physical risks cannot be dealt with in the same way, since if we told all individuals at risk of infection to find something else to do, we would be left with very few physicians indeed. Thus, even if physical and moral risks are equally real, moral risks might be avoided entirely, while physical risks can only be minimized.

Perhaps it is the case that social negotiation of the levels of risk that can appropriately be placed on physicians is relevant only for physical harms because would-be physicians unwilling to face moral risk should just not enter the profession. However, the consent model, our focus here, cannot get us to that conclusion because it is not at all clear that physicians with potential moral qualms have been expected to make such a significant sacrifice. Instead, we must ask whether physicians have consented to moral risk when they entered this profession or particular specialty, and if so, to how much? Just as physicians must know that they will face health risks in their job, they must also know that moral conflict might arise, but mere knowledge of moral risk, just as with physical risk, does not in itself indicate that physicians consent to accept all degrees of such peril, no matter how high. In fact, in many cases, "the core duties of a profession are never explicitly disclosed, and people in the profession may reasonably protest that they neither understood nor agreed to ... a constraint on their freedom [to follow their conscience]."[29]

Moreover, some physicians may have actually forgone promising professional opportunities in order to minimize the possibility of personal conflicts arising. Doctors may make conscious decisions about what specialties to choose and where to practice, both with regard to geography and institution, indicating that they did not actually consent to certain moral risks merely by virtue of becoming a physician. For example, a physician might choose to work at a Catholic hospital rather than at a public institution precisely to avoid certain moral conflicts, even if another hospital would have paid more, offered more prestige, or presented more interesting cases. Similarly, a physician could choose to practice in a morally homogeneous area where it is likely that the majority of patients would share the physician's moral views, or to practice in a heavily populated, diverse environment where many other physicians exist to render the service available even if the refuser has moral qualms. In both of these situations, one could reasonably argue that the refusing physician has not agreed to provide every service within the ambit of his specialty based on the fact that the objectionable service would either not be demanded or could be provided elsewhere. However, the clear rejoinder to this argument is that a physician's professional obligations are not up for personal negotiation and cannot be altered simply by where or what one chooses to practice.

Nonetheless, a physician's choice of specialty need not be categorized as a sort of personal negotiation. It could still entail the wholesale adoption of a set of professional obligations that simply differ between specialties; the physician is just selecting the preexisting set of corresponding obligations that is most in line with her moral views. Thus, a physician who would have preferred obstetrics might select neonatology instead in order to avoid having to prescribe contraceptives or perform abortions.

Unfortunately, however, even if physicians attempt to avoid specialties that pose obvious moral risks to them, as we saw previously, moral risks may not be static. Avoidance may become more and more difficult as technological advances within each specialty continually emerge and pose new hazards to conscience that were not on the horizon when the physician chose that field.[30] Similarly, what is accepted by general professional ethics may change over time, and what should we do about things that are legal but completely lack professional, or even public, consensus, such as physician-assisted suicide in Oregon? These changes may truly be beyond the scope of professional consent, but it is also quite possible that consent to provide even novel services can be implied, perhaps because drastic technological change is the very nature of modern medicine, rendering change itself predictable though the changes themselves may not be, or more likely, because the physician chose to remain in the profession subsequent to the change.

Regardless of this difficulty, none of these arguments based on lack of consent can absolve the physician who chose to specialize in obstetrics and gynecology and practice in a rural area with no other willing providers, entered the profession after the legalization of abortion, but nevertheless has strong moral objections to the provision of that service. Does that doctor have any plausible argument that, even under the constraints of the consent model, he is justified in his refusals to perform abortions? In fact, he does—and the argument successfully avoids the objection that individual refusers are inappropriately trying to alter the socially negotiated set of professional obligations.

A physician who was fully aware of the risk of moral conflict and who even failed to take mitigating action could nonetheless argue that he selected the profession in reliance on the widespread existence of refusal laws. And even physicians beginning their practice in or eventually moving to one of the few states that have not adopted such laws or whose laws do not cover the particular service the physician finds objectionable could at least claim reliance on the general historical understanding that physicians would have a great deal of freedom to select their patients. Importantly, the existence of conscience clause statutes themselves seems to evidence precisely the sort of negotiation and renegotiation of professional obligations in the face of changing circumstances that Daniels' consent model has in mind. If doctors ever were expected to provide services they found morally objectionable, there is at least some argument that such expectations are no longer reasonable.

Further, professional codes of ethics, which some commentators have called the most visible symbol of the moral foundation of the social contract between medicine and the community of patients,[31] almost universally recognize freedom of conscience for physicians without any of the service-based specificity often found in conscience clause statutes—and have consistently done so. For example, we saw earlier that the AMA's Code of Ethics supports a physician's prerogative to refuse to provide services that conflict with his personal religious, moral, or conscientious beliefs.[32] With regard to abortion specifically, the AMA's policy states that "the issue of support of or opposition to abortion is a matter for members of the AMA to decide individually, based on personal values or beliefs. The AMA will take no action which may be construed as an attempt to alter or influence the personal views of individual physicians regarding abortion procedures."[33] Similarly, the American College of Obstetricians and Gynecologists views physicians as "moral agents or decision makers, and, as such, [they] retain areas of free choice—as in the freedom not to provide medical care that they deem medically or ethically irresponsible."[34] Thus, like the AMA, ACOG allows its physicians to determine their individual positions on abortion based on their personal values or beliefs, though it has taken controversial steps to limit the scope of conscientious refusal in certain circumstances, albeit outside the group's official code of ethics.[35]

The fact that governments have had to provide incentives for physicians to render services in the face of epidemics indicates that physicians may not have consented to face all risks. By the same token, the fact that physicians have garnered such widespread protection of their interest in conscientious objection indicates that they did not consent, individually or as a group, to provide all legal medical services simply by entering the profession or a given specialty—they did not consent to accept all moral risks. Still, we must question whether this state of affairs is right; perhaps both situations are examples of physicians shirking their proper professional obligations.

While the consent model has the important virtues of refusing to unilaterally impose any obligations on individual physicians and preserving some semblance of predictability and stability across the profession, it is not entirely satisfactory in the context of determining whether physicians *ought* to be permitted to exercise their personal moral agency while donning their professional hats. The biggest concern with the consent paradigm seems to be that it is purely descriptive, perpetuating existing, often insular, professional norms. It offers no normative standard for assessing precisely which obligations should have been demanded of physicians in the past and should be demanded of them in the future, other than that we should at least demand that physicians consent to whatever obligations we decide on as a package deal.

Are the current professional norms asking too much or too little? Yes, we may have asked physicians in the past to accept the risk of personal infection, or we may have allowed them to avoid this risk when they would have been helpless to treat the infectious patient. And we seem to have allowed them to broadly reject moral risk associated with actions that would violate their personal conscience, but was this correct? The consent model cannot say—the norms are what they are, take them or leave them. In Daniels's ideal world, a patient could confront any specialist and expect exactly the same outcome. The problem is that the same outcome may simply be that each physician believes it is acceptable to refuse services on the basis of her own moral values. Thus, descriptive standards based on historical fact may be a good staring point, but they must be supplemented by more evaluative criteria in order to address the fundamental issue of patient access.

Another problem with the consent model is associated with its demand that physicians consent to the norms as they exist within the profession when the physician seeks entry, unless the physician is successful in procuring change across the profession as a whole. A doctor's personal values could conceivably change over time, and in fact might even be expected to, as the physician develops as a moral being and personally experiences things he had previously only heard about or never even imagined. Certainly, physicians should not be able to revise their professional duties in the midst of an existing doctor-patient relationship in which the patient has some reliance interest on the physician's past moral views and willingness to provide certain services. However, outside of these limits, and given the fact that a physician could establish the scope of her professional duties at least based on choice of specialty when she first became a physician, some types of revision may be acceptable.

For example, imagine a physician who was initially willing to provide nontherapeutic abortions, but who, after having children of her own, witnessing a particularly gruesome encounter with abortion, or experiencing a religious conversion, is no longer comfortable providing that service. If the fact that the physician initially chose a specialty that involves the provision of abortions is taken to mean that this particular physician must continue to provide abortions, the objecting physician must now change specialties or leave the profession of medicine entirely, forcing him to abandon significant investments in training and patient relationships as a result of a genuine moral transformation. Perhaps such dramatic measures would encourage physicians to cultivate deep levels of self-awareness prior to selecting a specialty, but the stability of responses to moral questions is not necessarily predictable. Given the drastic nature of forcing an objector out of his specialty and the fact that society could presumably benefit by retaining the physician, who is perfectly capable of providing other specialty services, interpreting the consent model of professionalism to accommodate at least some changes in moral values seems appropriate. An even easier solution to this problem becomes apparent, however, if a physician's socially negotiated responsibilities allow room for conscientious refusal, which, as explained above, they seem to. Society has simply not successfully negotiated a requirement that all ob-gyns perform abortion, so even if the doctor consented to that moral risk previously, his consent was not mandated and is entirely revocable.

As this discussion demonstrates, the consent model cannot convincingly be used to support the claim that physicians have a broad obligation to subrogate their personal moral beliefs in order to satisfy patient demands. Although the existence of the current debate might suggest that a renegotiation of professional responsibilities is in the works, for now at least, preservation of a physician's moral integrity does not appear to be a demonstration of illegitimate self-interest. Under this paradigm, physicians who choose to provide services that they believe to be morally reprehensible are in fact going above and beyond the call of duty.

The Patient-centric Paradigm

However, other views of medical professionalism reach quite the opposite conclusion. Rather than relying on what physicians have actually consented to as a descriptive matter, some commentators would hold them to a higher normative standard. The most extreme argue that physicians have an absolute obligation to provide all medical services falling within their professional specialty, or at least one that cannot be overcome by any moral objections a physician may harbor toward a particular service, regardless of the existence of more compromising alternatives. For these commentators, the baseline of what physicians must provide and the ceiling of what they can provide are one and the same.[36]

Proponents of this paradigm point out that the government spends roughly $110,000 per resident on training, rendering physicians deeply indebted to society and thus subject to its collective will.[37] They contend that if the moral stakes for refusers are so high, they should not risk conflicts at all and should simply do something else with their lives instead. Their forgone professional opportunity is considered adequately compensated by their resultant moral sanguinity.[38] More importantly, the argument goes, would-be physicians likely have enough other opportunities open to them such that there are no justice-based arguments that the "disability" or "supersensitivity" created by their moral beliefs must be accommodated.[39]

Since writing a short piece for the *British Medical Journal* on the topic of conscientious objection in medicine in 2006, Julian Savulescu has become recognized as a leading proponent of this patient-centric approach to medical professionalism.[40] He maintains that "to be a doctor is to be willing and able to offer appropriate medical interventions that are legal, beneficial, desired by the patient, and a part of a just healthcare system."[41] Savulescu demands that medical students be aware of, and ready to personally undertake, these deep obligations and argues that those individuals who are already physicians but who compromise patient access in order to pursue their own conscientious beliefs ought to be stripped of their licenses or punished through other legal mechanisms.[42] For Savulescu, a doctor's conscience has "little place in the delivery of modern medical care," and he goes on to assert that people who "are not prepared to offer legally permitted, efficient, and beneficial care to a patient because it conflicts with their values ... should not be doctors. Doctors should not offer partial medical services or partially discharge their obligations to care for their patients."[43] Thus, as a rule, he argues that the exercise of personal moral beliefs in a physician's professional capacity is an inappropriate breach of duty.

However, Savulescu's position may be less absolute than it first appears. First, it is imperative to note that his comments are directed at physicians practicing in countries with systems of socialized medicine, where patients can claim legal entitlements to medical services. He backs off substantially with regard to physicians in private medical systems, who he claims have more freedom to refuse patient requests that conflict with their own moral beliefs because they are not public servants.[44] Several American courts seem to draw a similar distinction, agreeing that the range of permissible conscientious refusals by government employees is severely limited and that they cannot delegate their obligations even to willing others.

For example, in *Shelton v. University of Medicine and Dentistry of New Jersey*, the court was faced with a nurse employed by a public hospital who chose to remain employed in the labor and delivery ward rather than accept a lateral transfer to a different department, but who nonetheless refused to participate in emergency abortions taking place on her floor. The court analogized to a case where a police officer who chose to remain employed in a district housing an abortion clinic was required to satisfy his assignment to stand guard outside of the clinic,[45] and stated that it would be "unremarkable that public protectors such as police and firefighters must be neutral in providing their services. We would include public health-care providers among such public protectors."[46] Thus, the court held that the nurse was obliged by her choice to remain on the labor and delivery ward to assist in abortions (and all other services provided there), just as the officer was obliged by his choice; the Church Amendment was not addressed. But for the legal exemptions often provided to doctors, a physician might be similarly forced to make a choice between specialties based on their covered services and to provide each of those services when asked.

Similarly, the Seventh Circuit in *Endres v. Indiana State Police* held that there was no obligation to accommodate a state trooper whose religious beliefs opposed gambling as immoral by reassigning him from a position as a Gaming Commission agent to a new post. The court concluded that law enforcement personnel cannot selectively choose which laws to enforce and who they will protect, even if allowing such selection would impose no hardship whatsoever. Instead, *each* officer must enforce *each* law.[47] Savulescu suggests that the same is true for physicians employed by the National Health Service—each one has a duty to personally provide each service falling within the domain of his or her specialty.

Although Savulescu is willing to draw a line between the obligations of public and private practitioners, others have argued that private physicians, because they have been granted unique privileges by the state, should actually be regarded as public servants.[48] Recall, however, that at least as a legal matter, licensing alone has been largely rejected as a basis for treating the licensee as a state actor. But even when a physician can be appropriately characterized as a public servant, such as when he or she is a government employee, it may be that Savulescu's argument goes too far.

Whether a person is a public servant or not, if no one is harmed by her refusal to provide some service because someone else is willing to offer it, it is unclear precisely why that refusal is impermissible.[49] Perhaps it is because the refuser would be shirking some obligation that is essential to her post. In fact, it is quite likely that police officers, for example, have a socially negotiated obligation to protect each and every person threatened by violations of the law, but we have already established that individual physicians have traditionally not been held to such high standards. Of course, that is merely a description of the status quo, and Savulescu and others believe that the standards for the medical profession should be raised.

Savulescu claims that when a physician has a "true duty" to provide care, "conscientious objection is wrong and immoral. Where there is a grave duty, it should be illegal." Unfortunately, he does not explain this seemingly important difference between true and grave duties, and a few interpretations are possible. One possibility is that the former exists when a physician is the only available provider of a service, while the latter arises when the patient is facing an emergency situation. However, because Savulescu did not directly address emergencies, it is unlikely that this distinction accurately captures his intentions.[50] The more logical inference based on Savulescu's other statements is that a physician has a true duty even if alternative providers are available and a grave duty when the physician's refusal would result in a complete denial of access.[51] Note that according to this approach, *all* refusers are unacceptably shirking their professional obligations, although Savulescu is willing to concede that not all obligations should be legally enforceable.

This interpretation of Savulescu's meaning is bolstered by his argument that the only possible justification for a physician validly refusing to provide a legal, efficient, beneficial medical service to a patient is some level of serious risk to the doctor's physical welfare. Even then, he asserts that mere self-preservation generally fails to justify a doctor's refusal to provide medical services. If that is the case, religious and other values, which pose no physical risks if violated, are entirely out of the running as legitimate reasons for failing to comply with a patient's request.[52]

Unfortunately, this view disregards the notion of moral risk and the stakes involved for refusers, or at least declines to give them any weight. Instead, Savulescu seems to believe that the responsibilities associated with being a physician trump all other responsibilities that a physician as a person might have, such as responsibilities to one's own moral integrity, one's family, one's religion, and the like. However, these other obligations are quite real and are widely recognized in other contexts. In fact, the person of principle is often held up as a societal ideal. A profession is only a single element of any individual's life and is not unique in having an impact on others—so do many other actions and choices taken and made in our personal lives. Thus, to expect, or even ask, physicians to give up all other considerations is far too demanding.

The patient-centric model is essentially a model of the professional as technician, complying with patient requests without significant (or any) room for personal autonomy or morality.[53] Under this complete rejection of medical paternalism, the doctor's sole function is to make up for the patient's lack of technical expertise and to resolve problems of information asymmetry, providing all of the facts truthfully and without judgment, leaving the patient to make the ultimate decision—and give the ultimate directive—in accord with his or her own values. Analogous arguments have been made with regard to the professional relationship between lawyers and clients,[54] although this approach has certainly not done the legal profession any public relations favors,[55] and has been heavily criticized in the medical context as eliminating the important caring role of physicians.[56] Nevertheless, this is precisely what Savulescu seems to be advocating. Such a dismissive response to the problems raised by conscientious refusal, however, inappropriately treats doctors as little more than simple yes-men to their patients.

Notably, Savulescu is willing to allow some leeway for the refuser when his or her values can be accommodated without compromising the quality, efficiency, or equitable delivery of a medical service desired by the patient. He suggests that when "many doctors are prepared to perform a procedure and known to be so, there is an argument for allowing a few to object out."[57] This exception, however, does not provide a sufficient standard for application. How many willing physicians would be adequate to allow a "few" to refuse, and how many would be more than a few? How might we deal with a situation in which there are too many objectors—should some of them be permitted to continue with their refusals (a "few"), while others are forced to provide the service, and how should we decide between them? For the reasons just described, and as we will see in more detail shortly, these concessions to physician conscience are necessary and appropriate, but it is insufficient to simply *excuse* refusers from a duty that remains intact under circumstances where the patient would truly be made no worse off. Instead, in that case, there is likely no individual professional duty at all.

Rosamond Rhodes rejects this claim, however, and takes a truly absolutist position, agreeing with Savulescu's overarching antipathy toward conscientious refusal in the medical context, but without making his concessions and without limiting her analysis to government doctors. Problematically equating moral beliefs with mere inclinations or preferences, Rhodes argues that when a person chooses to become a physician, he or she must cede authority "to professional judgment over personal preference."[58] She also takes the patient-autonomy movement to its zenith, analogizing physicians to soldiers, "obliged to follow the orders of higher ranking officers, who, in turn, must follow the direction of their political authorities."[59]

Rhodes claims that respecting physicians' conscientious objections unacceptably encumbers patients, and she accuses conscientious refusers of pure selfishness since they are willing to impose financial and other burdens on patients to avoid nothing more than their own "personal psychic distress."[60] She goes on to argue that just as lawyers and judges must follow the rule of law regardless of disagreement with it, personal conscience has no role to play in medical care, since this would undermine the trust patients place in doctors qua doctors, rather than qua individuals.[61] This analogy is imperfect, however, for it ignores the fact that judges can recuse themselves when their personal beliefs and biases are likely to impair their actions, and lawyers can generally exercise conscientious refusal if they would be unwilling or unable to zealously represent a client or present a particular argument.[62] Further, as will become clear in the next chapter, a physician's personal attributes, including his or her moral beliefs, can play a major role in the quality of the doctor-patient relationship; patients are often choosing a person, not just a professional. Why should their range of choice be unnecessarily limited?

Given that Rhodes does not assess any potential compromises to the conscience clause debate, we can infer that under her position, any level of patient inconvenience trumps the burdens imposed on an objecting physician who is forced either to provide services to which he or she is morally opposed or leave the field. Other commentators have adopted a somewhat different approach, one more sympathetic to physicians and less exclusively focused on patient interests than that taken by either Savulescu or Rhodes, but one that ultimately reaches the same conclusion. For example, Blustein and Fleischman explicitly recognize the significance of physician integrity and the inherent impossibility of asking physicians to separate their personal values from their professional roles. They note that "physician integrity is a value of fundamental importance in the practice of medicine. . . . If the physician were only to attend to the technical aspects of professional behavior, he or she would be no more than 'an engineer, a plumber making repairs, connecting tubes and flushing out clogged systems, with no questions asked.' This is a deeply unsatisfying and disturbing conception of the physician's role . . . . By dissociating from personal values in this way—at least those values that reflect what one takes to be most important—one only succeeds in creating deep divisions within the self."[63] Because they recognize these serious problems associated with forcing physicians to violate their consciences, Blustein and Fleischman conclude that conscientious objectors should not practice in certain specialties, not only out of a concern for patients, but also as a matter of protecting the physicians themselves.

At first glance, this line of argument does not seem objectionable. Certainly, if a person cannot fulfill the requirements of a job due to moral opposition to those requirements, she has no valid argument that she should be permitted to have that job. However, the fundamental problem raised throughout this section, and indeed, throughout this book, is how to define a job's requirements or to determine which services are truly integral to a professional specialty. For example, as we saw previously, despite efforts to protect patient access and restrict the exercise of conscientious refusals, the American College of Obstetricians and Gynecologists "respects the need and responsibility of its members to determine their individual positions [on abortion] based on personal values or beliefs."[64] This indicates that the provision of abortion services whenever those services are legally permitted and requested by a patient may not be an integral component of being an obstetrician or gynecologist. On the other hand, Blustein and Fleischman point out that ACOG's stance fails to acknowledge that "a pro-life position is not easily (in a moral sense) reconciled with the normal practices and procedures of maternal-fetal medicine," suggesting instead that willingness to perform abortions or the services leading up to it, such as genetic tests, actually is integral to the specialty.[65]

When consideration is given to the fact that a maternal-fetal specialist morally opposed to abortion could still provide excellent patient care in other aspects of the specialty, however, or that a fertility specialist could be an excellent physician even while refusing to provide certain types of assisted reproductive technology, and an oncologist could cure many patients of cancer even if refusing to provide those he cannot help with prescriptions to be used for a "clean" suicide, there is good reason not to cast too wide a net with respect to the services each physician in a particular specialty must be willing to provide—and certainly not to define essential services as all those that are legal. If we do take things too far in this regard, we are likely to run into problems when confronted with physicians who refuse to provide one or more services defined as integral, but nevertheless provide many other important and valuable services to their patients.[66] These refusers would essentially be sub-specialists, which is not inherently problematic, and those refused services would not be integral to their new branch of the profession.

This is not to say that we ought to allow physicians to entirely negotiate their own terms of responsibility, and there are certainly some baseline services that are integral to being a physician practicing in any specialty, which we will explore in chapter 9. However, these are likely quite minimal. Many opponents of conscientious refusal have simply gone too far, arguing for enforcement of an expansive and extreme duty that has not, need not, and as we will see, should not, be bargained for. Unfortunately, some advocates of conscience clauses have similarly overstated their claims, trading unyielding protection of patients for unyielding protection of physicians.

The Physician-centric Paradigm

At the opposite end of the spectrum from commentators leaving little or no room for the personal autonomy and moral integrity of physicians are those who argue that the "right" of conscientious objection ought to be inalienable and almost completely unrestricted even for professionals. These commentators take the conclusion reached by the consent model and push it to the extremes on normative rather than descriptive grounds. This distinction is critical, for while the very existence of conscientious refusal laws seems to outwardly suggest that society sees doctors as more than technicians who must put their special skills to whatever use is demanded of them, the fact that these clauses have been historically linked to abortion, contraception, and sterilization cannot go unnoticed. Rather than being a statement that conscientious refusal is acceptable, they may instead be a more accurate reflection of backhanded efforts to limit the availability of these controversial services, with physicians standing as nothing more than incidental beneficiaries. If conscience clauses are to remain, however, a more appropriate justification for their existence must be sought, and that is precisely what Edmund Pellegrino, chair of President George W. Bush's Council on Bioethics, and other advocates of permissive conscientious refusal policies aim to offer.

Before discussing Pellegrino's position, however, it is crucial to note that any attempt to keep religion and medicine entirely separate will likely prove impossible. This is not only because the great majority of American physicians profess religious affiliation, but more importantly, because theological claims have, throughout history, played a major role in defining a variety of concepts intrinsic to medicine, including life, death, health, and suffering.[67] Further, to ask that professionals leave their personal beliefs at the office door would be to demand that they become moral schizophrenics, which is psychologically intolerable given the inseparability of conscience from all aspects of one's life, including one's work.

A person's ethics are not instrumental tools that are capable of being turned on and off, implemented or ignored depending on the situation,[68] and even if they could be, doing so would pose serious problems for the doctor's moral integrity that should not be taken lightly.[69] This is because a physician's professional status does not mitigate her responsibilities as a moral agent bearing personal moral responsibility for her actions, regardless of whether they were requested by another and even if they were approved by society more generally.[70] Choices made in one's professional life inevitably have moral consequences for one's personal life. Therefore, it may not always be the case that patients have more to lose from a physician's refusal than the physician has to lose from acting against her conscience—but then again, this is precisely why commentators like Blustein and Fleischman suggest that would-be refusers simply avoid the profession.

This suggestion is quite problematic from Pellegrino's perspective, however, since, as he points out, nearly all clinical decisions involve value judgments of one kind or another. Because physicians must frequently make moral choices and regularly do so according to their own conscience, value neutrality is a logical impossibility.[71] Thus, he argues, demanding it of physicians and excluding only those whose values are based on religion or other personal moral beliefs that conflict with those held by the majority reflects nothing more than an unacceptable preference for one set of values as the only acceptable values in a pluralistic society.[72] Others have agreed that this is simply "bad-faith authoritarianism, a dishonest way of advancing a moral view by pretending to have no moral view."[73]

Pellegrino notes with concern that the growing focus on patient autonomy has come at the expense of physician autonomy, and worries that this will result in physician freedoms being inappropriately trampled. He recognizes that patient autonomy arguments may be compelling when they address freedom from coerced medical care (i.e., negative rights to be left alone and to refuse unwanted treatment), but suggests that they are far less persuasive as a reason to compel another to act affirmatively, since that other person will also have autonomy rights and an interest in preserving his own moral integrity.[74] Allowing the claims of patients to categorically trump the claims of those who choose to become physicians is unwarranted, Pellegrino claims, because "autonomy gets its status as a moral right of humans from the fact that human beings have the capacity to make rational judgments about their own lives, choices, and interests.... To obstruct the capacity for autonomy is to assault an essential part of a person's humanity, because the choices we make are so much an expression of our membership in the human community, of who we are or what we want to be as individual members of that community."[75] In other words, doctors are people, too.

This argument is somewhat less convincing than Pellegrino's first claim alleging discrimination against, or at least inconsistency toward, conscientious refusers, since it fails to explain why all restrictions of autonomy that prevent harm to others should not similarly be rejected, though that clearly cannot be the case. Nonetheless, based on the view that autonomy is an unwaivable human right applicable to all persons regardless of role, Pellegrino and others conclude that physicians need not sacrifice their freedom of thoughts, words, and importantly, actions simply as a result of their entry into the profession.[76] Instead, Pellegrino sees the preservation of a physician's personal moral integrity as an example of a legitimate self-interest that ought not be subrogated even to what the patient may believe is a morally good, or at least acceptable, purpose.[77] Charles Fried has drawn similar conclusions with regard to the legal profession: "Just as the principle of liberty leaves one morally free to choose a profession according to inclination, *so within the profession it leaves one free to organize his life according to inclination.* The lawyer's liberty—moral liberty—to take up what kind of practice he chooses and to take up or decline what clients he will is an aspect of the moral liberty of self to enter into personal relations freely."[78]

That there ought to be *room* for the personal moral autonomy of physicians even in the professional setting and that patient demands need not *always* win out are relatively moderate and compelling propositions. Even if the moral right to exercise one's conscience is not absolute and can be challenged by one's professional obligations, freedom of conscience is of such great consequence that we should not demand waiver unless and until other options for accommodating patients' interests have failed. The notion that doctors and their patients are on completely equal footing in all contexts, however, can be quite troubling.

If patients should not be allowed to impose their views on physicians, then physicians should be prohibited from doing the same to their patients, for both hold the moral right to autonomy. In many situations, physician refusers can avoid that result, but when a true conflict arises, something has to give. Pellegrino and certain other extreme proponents of conscience clause protection argue that patient access is that something. In their eagerness to protect personal autonomy for all, however, they end up problematically placing physician freedoms ahead of the interests of those relying on them for care.

For example, Pellegrino has consistently been unwilling to modify his understanding of the physician's professional obligations even when refusal creates significant access problems for the patient. When no other physician is available due to geographic restrictions, clinical urgency, or competency constraints, but the moral differences of the physically available doctor and patient remain irreconcilable, he has argued again and again that the physician must remain faithful to her conscience, impact on the patient notwithstanding.[79] Rhodes has rightfully criticized Pellegrino for the uncompromising nature of this version of medical professionalism, indignant at his refusal to recognize professional obligations sufficient to override personal values in these "hard case" scenarios, despite acceptance of this ranking in other fields.[80] For example, even those commentators eager to protect the moral autonomy of professionals are generally willing to limit that autonomy in the last-lawyer-in-town cases, acknowledging that there is some moral duty to represent a client in need of legal services within one's competence who cannot otherwise find counsel, moral disagreement notwithstanding.[81] Nevertheless, many state conscience clauses seem to at least partially accept Pellegrino's claims, perhaps revoking protection in emergency situations, but generally retaining it without regard to whether there is an alternative provider reasonably available to the patient.

Pellegrino argues correctly that physicians have the right, and perhaps the obligation, to stand up for their personal beliefs in the public square by trying to persuade others through political action and public debate. He also argues that the physician's moral claim to free exercise of conscience allows her to individually refuse to provide medical services she finds morally objectionable, avoid accommodation of secular beliefs, and resist the changing beliefs of other physicians[82]—and normally, boycotts or protests are an acceptable way of bringing one's views to the fore and potentially changing the status quo. However, given the state-granted monopoly power of the medical profession, if enough physicians refuse to provide a given service, it will become effectively inaccessible to patients. In such a circumstances, physicians would be exercising quasi-legislative powers without equivalent democratic legitimacy. Their boycotts could be unduly powerful and might result in an inappropriate imposition of their views on patients. In fact, this is precisely what opponents of conscientious refusal are concerned about, both in this country and abroad. For example, both the United States and Britain are facing an abortion crisis as an unprecedented number of physicians refuse to participate in the procedure, leading to concern that antiabortion groups will succeed in achieving their goal not through means of legitimate debate, but rather by default.[83]

This is quite troublesome because, as Pellegrino himself recognizes, "decisions that must be made as a matter of public policy in areas such as abortion, euthanasia, and stem cell research, should be made democratically, universally, and equally binding."[84] If this is the case, then objectors should not be allowed to unilaterally change public policy decisions for individual patients by impeding the availability of medical services. Importantly, however, no silencing of opposing viewpoints occurs so long as the patient can reasonably access the desired service from someone. The proper solution to the conscience clause controversy takes account of this fact by conceptualizing the responsibilities of individual physicians as independent from the responsibilities of their profession as a whole.

The Gatekeeper Paradigm

The best, most pragmatic, and most broadly acceptable version of medical professionalism for the purposes of the conscience clause debate draws physician responsibilities from their position as social gatekeepers. Robert Goodin offers perhaps the clearest and most compelling account of the profession's obligation, though his is intended to be a broader theory of social responsibilities not applicable exclusively to doctors. Rejecting the idea that special duties result solely from their voluntary assumption, Goodin claims that it is dependency and vulnerability that give rise to our most fundamental moral obligations. He argues that a person's responsibilities are based on the extent to which others are reliant on him or her alone to perform certain services and the extent to which others are vulnerable to that person's actions and choices; in other words, you have a duty to help when you have a special ability to do so.[85] The medical profession's collective monopoly power renders members of the public uniquely vulnerable to and dependent on the profession for access to services falling under its monopoly, and so under Goodin's theory, the profession bears special obligations to relieve and ease that vulnerability by making sure those services are readily available.

Notably, however, reliance on monopoly power and the related defenselessness it creates may generate duties greater than many are willing to accept. Critics could easily reject Goodin's model as proving too much, since taken to its logical conclusion, it would impose an obligation on companies with pharmaceutical patents to ensure that their products are accessible to those who need them, or even on a sole onlooker to rescue a drowning stranger because he is in a unique position to do so. Our laws currently enforce no such responsibilities, though many believe they should, or at the very least, that moral duties are present. Nevertheless, for those who believe this goes to far, we can point to a couple of potentially relevant distinctions—limitations on Goodin's otherwise broad imposition of duty. First, pharmaceutical companies are businesses and the patent system trades access for innovation, while bystanders have undertaken no voluntary commitment and enjoy no special relationship with the victim. More importantly, neither represents a "profession," but professions have heightened responsibilities both as a matter of virtue and as a result of their socially negotiated power, which establishes special relationships with the public.

In fact, a defining characteristic of all professional groups is that, for a variety of reasons, they have been granted "a corner on some valuable form of knowledge within a society. Wherever this is the case, there is power—power to control the knowledge itself, and especially, power over the aspects of human life that depend upon this knowledge."[86] The community assures itself that professional power will not be abused by instituting professional obligations, reciprocally trading special privileges for special duties to use those privileges for the good of individuals and society.[87] Thus, the medical profession's exclusive power over most mainstream medical services, combined with the commitment of all professions to altruism and renunciation of pure self-interest, are elements that work together to establish the profession's duty to ensure that the conscientious refusal of individual physicians does not impose significant barriers to access for patients. Even if there have been challenges to the profession's power as a consumer model of medicine has become more deeply entrenched, its collective monopoly power—and associated duties—remain.

Again, however, we may have proved too much, since this combination of factors seems to also suggest that the profession has an obligation to mitigate *all* potential barriers to patient access, particularly financial barriers. Of course, members of the profession are entitled to expect remuneration for their services, obligations to provide charity care notwithstanding, and they have no obligation to operate at a loss.[88] Nevertheless, despite the fact that physicians are allowed to profit from their professional endeavors, the profession may indeed bear an obligation to ensure that the services within its collective monopoly are financially obtainable by those who need or desire them. This is certainly the case according to Goodin's analysis.[89] However, this responsibility may also be shared with others beyond the profession, namely government health policymakers, who have the power to resolve these problems as well, through insurance programs and subsidies, for example.

Similarly, if patient access is at least partially the profession's duty, we might be powerless to avoid elimination of other traditional aspects of physician autonomy, such as their ability to control who they treat, where they practice, and what they practice, given that these freedoms have had clear negative effects on access to care by important segments of the population. In fact, this is precisely what Daniels has argued, stating that the scope and content of the profession's ethics should be determined by the need to guarantee the just distribution of medical goods and suggesting that the individual autonomy of physicians could be appropriately limited in scope in order to accomplish these distributive goals.[90] However, these constraints would be unnecessary, and thus uncalled for, if other freedom-preserving alternatives—such as positive incentives for some choices as opposed to negative restrictions of others—could achieve the desired result, which they can. More importantly, we will see that such constraints ought to be avoided in all but the most extreme situations given their potential to do more harm than good.

While barriers to access created by factors other than conscientious refusal are of the utmost importance, particularly since things like the inability to pay for care and problems associated with rural medicine are likely to affect far more people than a physician's moral unwillingness to provide a service, they are beyond the scope of this book's narrow call to resolve the problems raised by conflicts of conscience. It is not inherently troubling to say that the profession has a responsibility to work toward solutions for all sorts of patient access problems stemming from patients' exclusive reliance on the profession for medical aid, so we have not actually proved too much. Instead, we are legitimately selecting one barrier to access that has become hugely controversial and focusing exclusively on that.

Thus, given that preventing conscientious refusal from burdening patient access is at least one of the profession's obligations, the crucial question becomes whether the clear duty held by the abstract, collective entity of the medical profession is also held *personally by individual physicians* regardless of context. Numerous commentators have responded to that question in the negative. For example, R. Alta Charo has noted that "by granting a monopoly, [the states] turn the profession into a kind of medical utility, obligated to provide service to all who seek it," but she goes on to state that "accepting a collective obligation does not mean that all members of the profession are forced to violate their own consciences." Instead, it means only that a system must be implemented to ensure that patients can exercise their autonomy as readily as physicians can.[91] Similarly, Rebecca Dresser has stated that "although every individual member of the medical profession might not have a duty to perform every procedure within his or her competence, society expects the broader profession to adopt reasonable measures to promote patient access to medically acceptable procedures."[92] And finally, Daniels has recognized that access is a social obligation that "does not individuate directly into an obligation on each physician or provider to deliver that care."[93]

Importantly, Goodin's vulnerability model reaches the same conclusion as each of these commentators. Recall that the strength of the actor's responsibility depends strictly on the degree of the other's vulnerability to the actor's choices and behavior—the "only way to avoid the responsibilities would be to eliminate the vulnerabilities."[94] Thus, the actor's responsibility is mitigated and potentially eliminated when the other is able to find alternative sources of assistance or protection, precisely because the magnitude of potential harms the other might suffer consequent to the actor's choice to refuse assistance is diminished.[95]

While the exclusivity element of professional power has been used to support arguments that individual physicians cannot justifiably deny care on grounds of conscience, it is essential to recognize that this factor is really applicable only to the profession as a whole, or in the rare case, to the "last doctor in town." Although the profession *collectively* wields monopoly gatekeeping power, individual physicians are not appropriately described as monopolists when patients have reasonable access to other willing providers. In this circumstance, patients are not uniquely vulnerable to the whims of only a single physician, so with the vulnerability eliminated, refusers can appropriately avoid responsibilities that might otherwise exist to provide services they find morally objectionable—*note that they truly lack a duty, it is not simply excused.* True, they could open the gate to the service if they so chose, but a patient desiring access could also pass through the gate if it were opened by another physician. The profession's obligation is fulfilled even if not every individual physician participates,[96] although we will see that there are still some minimum obligations that must apply to individual doctors.

Thus, the gatekeeper/vulnerability argument and its associated duties retain force only when applied to the profession collectively or to circumstances in which the individual physician stands in the place of the collective profession due to a patient's inability to access another willing provider. Physicians finding themselves in this latter situation cannot appropriately refuse their services on moral grounds, since it is impermissible to claim an absolute right to personal autonomy when others are forced to rely on you for access to a public good.[97] However, it is essential to recognize that the circumstance of the lone physician holding the profession's control will be relatively infrequent, usually occurring only in rural areas or highly specialized fields of medicine. Further, even these infrequent scenarios can often—and should—be avoided through institutional schemes explored in subsequent chapters.

Of course, opponents of conscience clause protection would refute these claims, likely arguing that effacement of self-interest is a professional obligation held by each individual member. However, we have seen that the precise level of self-denial expected of physicians is hotly contested. More importantly, this gatekeeper model of professionalism, which bifurcates the obligations of the profession and professional in most circumstances, is the only approach capable of laying a solid foundation for the maximum protection of both doctors and patients. Thus, even if we set aside monopoly power and disregard what has been expected of professionals in the past, it would still make sense as a consequential matter to impose the burdens of access on the profession collectively.

This solution may be difficult to implement, but it is our best bet for moving things forward.[98] It will require naming a tangible institution to stand in as an appropriate proxy for the abstract concept of the collective profession, and will also entail the development of mechanisms that will allow that proxy institution to carry out the profession's obligations in a way that does not force doctors to sacrifice their moral autonomy except when absolutely necessary. These difficulties may explain why no compromise has yet gotten off the ground in the context of physician refusals, but resolving them is not beyond our reach and we have every reason to work steadily toward achieving this goal.

Each of the models of professionalism we have explored in this chapter offers an important perspective on the conscience clause debate, but only one of them can get us out of the current stalemate and move toward a true balancing of interests. Unfortunately, many commentators seem to take the primacy principle either too far or not far enough, without recognizing that protection of the interests of both doctors and patients rarely needs to be mutually exclusive. Forcing all objectors out of medical specialties in which conflicts of conscience might occur is sufficiently harsh as to demand exploration of alternative models before adopting that solution, but allowing such refusers to remain in the profession without resolving the problems their refusals can pose for patients is not appropriate either. Overly broad protection of conscientious refusal could do a great deal of damage to the profession, resulting in a lack of any professional obligations whatsoever since any and all could be trumped by a physician's personal beliefs.

In short, doctors ought to be asked to do more, but not to do everything. Distinguishing between the responsibilities of doctors as individuals and doctors as a group allows us to reach the proper middle ground. Given the importance of personal moral integrity and adherence to individual conscience, it seems likely that citizens forming a social contract with the profession of medicine would agree that physicians can refuse to provide objectionable services so long as access to those services is still reasonably available. While a focus on legal rights has plagued existing commentary on the matter of physician conscience, examining the moral duties of the profession has taken us much further. The remainder of our discussion will examine exactly how the profession can and should go about satisfying its obligations in the shadow of conflicts of conscience between doctors and their patients.

II

Protecting Doctors and Patients: An Institutional Solution

3

Moral Diversity in Medicine and the Ideal of Doctor-Patient Matching

Given that making medical services available to the public is a fundamental responsibility of the medical profession, we must ask how we can secure such availability in the context of the moral opposition felt by some members of that profession to some of the services patients may seek. One option might be to explore elimination of the collective professional monopoly held by physicians in many controversial areas. In fact, because physicians have failed to adequately provide access to some important services for women, a handful of states have decided to allow nurses and other trained health-care professionals to perform abortions and have not experienced any measured decline in the quality of care.[1] If the profession's obligations are based on its gatekeeping power, removing that monopoly would result in less power and therefore less responsibility, thus spreading the burden in the same way individual physicians usually can. However, this may not entirely eliminate the problems posed by conscientious refusal, since medical services will always need to remain under some sort of professional monopoly in order to protect patients. Expanding this collective monopoly to include other professions will help, but some professionals with moral objections—and the related potential for access problems—will inevitably remain.

Alternatively, we might allow physicians to conscientiously refuse so long as another physician is available to satisfy the patient's request, while requiring the objector to meet patient demands regardless of his moral qualms if no substitute can be reasonably found. To mitigate the burden on doctors, we could simultaneously attempt to eliminate situations in which individual physicians bear the monopoly powers of the profession as a whole. This solution avoids the extremes and will accommodate the moral integrity of physicians as well as the access interests of patients by adopting the sort of institutional model that has been successful in dealing with conscientious refusals in other areas of medical service, using a scalpel rather than a broadsword to resolve conflicts of conscience in health care.[2] Due to its capacity for compromise, we will see that the institutional solution is the most promising, though not without some problems of its own.

However, before delving into this alternative, we must appeal once more to those who may not have been persuaded by the critique of the patient-centric paradigm offered in the preceding chapter and who may still believe that the best approach is for the profession to satisfy its obligations by allowing entry only to those individuals willing to provide all legal medical services falling under the umbrella of their chosen specialty. Unfortunately, this seemingly persuasive logic seriously oversimplifies the problem and imposes an unnecessarily heavy and unrefined toll. It ignores the value of moral diversity within various medical specialties, fails to recognize that few medical services are so integral to a given specialty that it would be impossible to practice as a specialist while refusing to provide them, and overlooks the possibility that such strict exclusion of conscientious objectors could predictably result in widespread shortages that only exacerbate access problems for patients. As this chapter begins to explain, excluding refusers entirely from various fields of medicine would not only be detrimental to the objectors themselves, but also to the profession, patients, and, most importantly, society more generally. So long as patient access to the appropriate basket of medical services is protected, as it would be under the institutional approach explored in part III, there is an important role for conscientious refusers in medicine.

Rationales for Maintaining Physician Conscience

At a fundamental level, creating an automaton model of medicine, in which physicians must choose between continued dedication to the profession or their own personal ethical integrity, will leave them detached from potentially appropriate moral qualms and from their patients. This will certainly corrode the humanity and compassion patients expect and need from their doctors.[3] Worse still, imagine the morally insensitive people we would attract to the profession if we decline to preserve any role for moral discretion in medicine. Few physicians are willing to provide any service a patient requests so long as it has not been explicitly prohibited by the legislature—who might be left if we told doctors that this is in fact what is required of them?

Speaking in reference to the legal profession, Sanford Levinson has objected to what he sees as the predominant professional model that strives to create "almost purely fungible" members of the professional community.[4] This "bleaching out" of one's religion, moral base, race, gender, ethnicity, and culture by professional norms is problematic, especially if these differences among professionals from diverse backgrounds can actually improve the service provided to clients. Similarly, Howard Lesnick argues that a professional's desire to guide her actions by a sense of obligation to perceived moral imperatives should absolutely not be bleached out by the norms of professionalism. Instead, it should be "presumptively welcomed as [a] socially desirable characteristic likely to enhance the quality and consequences of interactions between professionals, their clients, and the systems in which they interact." He goes on to make a point that is worth quoting at length: "A polity that encourages its citizens to bring to bear their own serious moral reflections on the morally significant decisions they face will be more likely to grow in justice and humanity. In finding in its pluralist quality a counsel of restraint in its encounters with the varying religious scruples of individual attorneys, society is therefore not merely indulging an individual's interests, but recognizing their 'collective value' as well."[5]

This counsel of restraint is especially important in the medical context, where we must question to what extent and how quickly we should enter the brave new world offered by novel medical technologies, as well as the uses to which we should put existing ones. Some commentators have expressed fear that "the day is rapidly approaching when health-care providers who refuse to kill Grannie—or to give up on an imperfect child—will no longer exist in American medicine. And on that day, no clinic or hospital will ever again be a refuge against disease."[6] Hyperbole aside, having doctors who are permitted to refuse is a comfort against precisely this sort of concern.

Allowing room for conscientious refusal permits physicians to think harder about their actions and imposes a greater sense of moral accountability in the practice of medicine. As we saw previously, public distrust of the legal profession and the widespread perception of lawyers as hired guns, willing to twist the law to its limits in an effort to advance any and all client aims regardless of their moral propriety, appears to be one effect of the professional-as-technician model. Similar professional consequences might follow for medicine should physicians adopt an approach in which moral grounds for refusal are prohibited. Instead, we ought to recognize, as Lesnick has, that preservation of some level of conscientious refusal and moral diversity within the profession is beneficial not only to refusing physicians, but also to the community as a whole.

The counsel of restraint rears its head in other ways as well. First, we need only look back over the course of a few short decades to recall the consequences of a medical profession unable or unwilling to object on grounds of conscience to morally abhorrent acts. Consider the Tuskegee study in which physicians conducted research on the natural course of syphilis using unwitting African American men as subjects even after the introduction of penicillin as the standard of care,[7] and the Willowbrook tragedy in which doctors intentionally infected institutionalized retarded children with hepatitis in order to research the disease's natural course and preventive measures.[8] Consider also the well-publicized medical experimentation on prisoners of the Holocaust, conducted by physicians who had been otherwise described by the inmates of the concentration camps themselves as "kind" and "decent," and who in their personal lives enjoyed reputations as caring and sensitive, but who nonetheless committed atrocities in the discharge of their public functions.[9] Finally, consider the participation of today's physicians in military torture[10] and state-sponsored executions.[11]

Granted, these examples deal with physicians doing things to patients that those patients did not want, or would not have wanted had they been aware of the circumstances, and this is a major reason that they have been historically and professionally condemned. Conscientious refusals, on the other hand, generally involve patients actively and competently requesting services for themselves and nevertheless being denied. However, some of the most frequent circumstances of conscientious refusal involve patients asking for things that affect others, like society more broadly or their would-be children. For example, if we controversially assume that the fetus is entitled to the same level of respect as a full human person, or less controversially that it is at least entitled to substantial moral weight, the mere fact that a patient asks for an abortion does not make it per se morally right; the fetus did not, and likely would not, agree to the request. Further, even if we believe that there is no harm in not coming into existence, but rather just lack of benefit, the fetus as a person or as a being of moral worth has already come into existence. Thus, the fetus is more than simply not benefited by abortion, but instead is actually harmed. Thus, the analogy to the parade of horribles listed above begins to appear more compelling.

In fact, there may even be entirely self-regarding medical procedures requested by competent patients that it is valuable for physicians to stand up against, and it is important to see that the consent of some party does not completely eliminate any value in physician restraint. First, not everyone accepts the notion that competent people should be permitted to hurt themselves, for self-harm may be wrong as inherently violative of human dignity, perhaps as in the case of female circumcision. Or it may be wrong as a result of its effect on future choices, which is why even strong liberals reject the freedom to sell oneself into slavery.[12] This argument could logically be extended to the refusal of life-sustaining care, physician-assisted suicide, and voluntary euthanasia since each of these things results in death and thus requires a relinquishment of future autonomy.

Further, even if self-harm should be allowed, it is not the case that it should be encouraged. Others likely ought to take measures to actively discourage it through conviction and persuasion, even if not by compulsion. As John Stuart Mill explains in his famous essay "On Liberty," "human beings owe to each other help to distinguish the better from the worse, and encouragement to choose the former and avoid the latter. They should be forever stimulating each other to increased exercise of their higher faculties, and increased direction of their feelings and aims toward wise instead of foolish, elevating instead of degrading, objections and contemplations."[13]

Of course, it is true that a doctor willing to perform the controversial procedure for a patient may nonetheless be able to adequately articulate the moral case against it. For example, South Dakota enacted a law that requires physicians to inform women seeking an abortion that the procedure would "terminate the life of a whole, separate, unique, living human being," that the woman has a relationship with the unborn child, and that there are alternatives, such as adoption. The statute also requires doctors to inform women of abortion's medical risks, including psychological effects on the mother.[14] Unsurprisingly, this law has been challenged and, at the time this book went to press, was under en banc review by the Eighth Circuit Court of Appeals after being struck down by a panel of three judges from that court.[15] Nevertheless, it demonstrates that perhaps we do not actually need refusing physicians to offer the counsel of restraint, though it is likely that the arguments will be made best by those who are actually convinced by them.

Aside from simply offering persuasion and information, however, physicians might appropriately say to patients who cannot be convinced to act otherwise, "hurt yourself if you must, but I will not help you." If the behavior in question is truly wrong, that is the proper response and it is important to safeguard physicians who will react in this way. But this matter is quite complicated, as we will see in a moment, given that in many cases we cannot be sure that the behavior *is* truly wrong, and if a physician will not help, the patient may be precluded from doing what he or she thinks is right or permissible. The bottom line here is that while patient consent can significantly alter the situation, it does not necessarily create a moral free-for-all, for it is possible that what the patient wants is ethically wrong even if there are insufficient grounds to prohibit it outright.[16] The request itself, combined with the imprimatur of legality or even general public or professional acceptance, is not the equivalent of a stamp of moral correctness.[17]

Building on these important considerations, the driving force and strongest argument for retaining room for moral refusers in the profession is the fact that many of the issues facing physicians raise metaphysical questions entirely immune to empirical testing or any other comprehensive doctrine for distinguishing right from wrong. The answers are simply not knowable, so we do our best to choose correctly, but one group rarely has the authority to tell another that its perspective is wrong.[18] Thus, as Mill has persuasively explained, we benefit from maintaining diverse viewpoints, excluding only arguments that are entirely illogical, for the ensuing debate will help siphon out the most accurate version of moral truth as errors are revealed and persuasive arguments are strengthened through their collision with error.[19] Again, these diverse viewpoints may not all need to be represented within the profession itself, for the profession's behavior can be influenced by wider social discussion, but there is still an important reason to maintain a protected role for objecting physicians. If we cannot be completely sure that we have gotten it right with regard to any of the legal or professionally sanctioned medical services to which physicians currently express moral opposition, *there is a distinct possibility that the refusers are right*, leaving no legitimate grounds on which to exclude them from the profession. The problem is that we often just do not know.

For example, the federal government prohibits the performance of female circumcision on minors, and several states extend the prohibition to adult women,[20] but does a physician's refusal to perform that procedure where it is illegal become inappropriate where it is legal? If female genital cutting really is a human rights violation, legality of the procedure should not matter, but if it is simply a matter of cultural differences, perhaps an adult woman who is fully aware of the personal risks should be allowed to accept those risks for herself.[21] Similarly, "what are we going to say when (no longer if) active euthanasia becomes legalized and comes to be viewed as a right to which patients are entitled, much like what already seems to have happened regarding the right of patients to have life-sustaining treatment withdrawn? Will we, a decade or so down the line, be considering removing the privileges of those who not only refuse to perform it, but refuse to participate in such an enterprise in any fashion? ... But then we are talking about the actual killing of human beings, and a profound moral intolerance of others. And *that is precisely* what our true conscientious objector is maintaining now, regarding the current issue of withdrawal of treatment," and abortion as well.[22] Clearly, moral uncertainty in medicine abounds.

In this context, to make outcasts out of "moral whistle-blowers" carries no societal benefit.[23] In fact, some commentators seem to rely on similar logic when they suggest that in areas where medical and political communities lack consensus,[24] physicians should not be forced to provide the service in question, especially due to the potentially "sobering permanence" of their actions.[25] Unfortunately, this conclusion begs the question since it would require some notion of how much consensus is sufficient for us to feel confident that we know the right answer, or at least for moral refusals to no longer serve a valuable function. Nevertheless, this approach would at least allow the state to avoid "arbitrating metaphysics," permitting both doctors and patients to do what they think is right when we are unsure about what is wrong.[26]

Preserving room for *both* sets of views is crucial, however, for as we saw in the context of the physician-centric model of medical professionalism, just as the beliefs of objecting physicians should not be unnecessarily suppressed, objecting physicians should not be permitted to suppress the beliefs of others. This is a serious risk given the gatekeeping power of the medical profession, which itself could be used to silence opponents who have no alternative access to a refused service and whose voices are equally important to the debate. As Mill notes, "we have a right ... to act upon our unfavorable opinion of any one, *not to the oppression of his individuality*, but in the exercise of ours."[27] Thus, we must avoid taking the counsel of restraint argument too far. Instead, we must take steps, the details of which are the focus of part III, to ensure that patients are not faced with a situation in which every available physician refuses to satisfy a request that is only arguably immoral. In that case, we would lose the value of moral diversity and would instead have an illegitimate imposition of views without any assurance that those views are in fact correct.

While many have characterized moral refusers as overtly selfish, placing their own interests above those of their patients, on closer examination, we see that there are a variety of reasons that it is important to maintain room for conscientious objection by physicians. Society benefits from having morally serious people in the profession who are unwilling to just follow orders and who contribute to the rich moral debate that helps avoid blindly accepting the normative permissibility of whatever is technically possible and has not been legally prohibited. Even if one remains unconvinced of these broader benefits, however, all is not lost, for it is certainly true that physicians themselves benefit from avoiding the internal turmoil stirred up when forced to violate their conscience—

these are often things that they *cannot* do, not just that they prefer to avoid. Compared with the relatively small inconvenience patients are asked to bear in finding an alternative provider, particularly when aided by all of the mechanisms and guarantees explored subsequently, the benefit to physicians standing on its own is sufficient to justify the preservation of some room for refusals on grounds of individual conscience. In fact, patients may find that what at first appeared inconvenient is actually an opportunity to preserve and augment their own autonomy through selection of a physician who shares one's own deeply held moral beliefs and who might have otherwise been forced out of the profession.

The Value of Doctor-Patient Morals Matching

Fortunately, there is a more palatable solution to the conscience clause dilemma than exclusion of particular moral perspectives, one that recognizes the value of including moral objectors in medicine so long as patients can access desired services from some source. This solution recognizes that professional excellence can present itself in multiple forms and that physicians can appropriately select different subroles expressive of a wide variety of moral virtues, establishing a further division of labor within the profession.[28] There may in fact be as many strains of medicine as there are value systems, and patients may prefer to be treated by physicians with similar values, who, while refusing to offer certain medical services normally within the ambit of their chosen medical specialty, are nevertheless capable of providing competent and valuable patient care. Rather than banning or sanctioning these physicians, the government, the profession, and society should allow both doctors and patients to seek out like-minded others. This will not only protect the autonomy of both parties in this era of religious and moral pluralism, but will also avoid a situation in which physicians who lose the public debate with regard to moral permissibility "are precluded even from maintaining a subcommunity that enables them to live out their convictions."[29]

There is no reason that all physicians or specialists must each agree to provide precisely the same basket of services, and such a requirement has never been in place. In fact, there is already significant latitude for doctors and patients to negotiate their own relationships and set ground rules for acceptable actions within them. Physicians can and do limit the scope of their practice to a particular specialty and geographic area, and even beyond that, may limit themselves within their specialties, such as the pediatric endocrinologist who focuses exclusively on children with gender identity disorder or the orthopedic surgeon who only performs hip replacements. Analogously, there should be no problem with allowing doctors to further limit their practice based on their moral beliefs as a type of subspecialization, and this is particularly the case if patients see some value in this approach.

This leads us to precisely the sort of matching model—or deep-value pairing of professional and patron—that has been strongly advocated by bioethicist Robert Veatch as the future of medicine, as well as by David Orentlicher as a way of ensuring that a patient's end-of-life wishes are satisfied to the fullest extent possible.[30] It has the notable advantage of preventing either professionals or laypersons from being "forced into the intolerable situation of having to choose between violating one's conscience and violating the conscience of the other party."[31]

Veatch envisions a system in which patients and physicians establish their professional relationship only after exploring the basic moral commitments held by each and determining that there are unlikely to be any insurmountable moral disagreements about services that would be expected to come up during the course of that relationship.[32] This may appear to be unbearably time consuming in an era in which physicians are already extraordinarily pressured in the number of minutes they can spend with patients. However, there are a number of reasons, conscientious refusals aside, to develop a mechanism to finance health-care services in a way that will properly reimburse, rather than disincentivize, physicians for taking the time to discuss values with their patients, and so this concern should not be treated as fatal to Veatch's proposal.

He suggests an arrangement with several levels of contracting for professional services that is quite similar to Daniels' consent model. The broadest involves the basic social contract, followed by a contract between the lay population and the professional group, determining the baseline of core services that are integral to the profession and thus cannot be avoided by any member.[33] Finally, Veatch would allow individual patient and physician pairs to contract for whatever was mutually acceptable to each of them, so long as this personal contract did not violate any provisions of the broader contracts between the profession and society, which serve as constraints.[34]

As discussed previously, the fact that the profession of medicine has been given a collective monopoly on the provision of certain medical services indicates that the public would likely demand that the profession make the basket of services falling under its control reasonably accessible to patients. If individual physicians can contract around the provision of services they find morally objectionable, as they should be allowed to, the overarching profession would have to guarantee sufficient diversity of physicians such that appropriate contractual matches are possible and there is no truncation of styles from which patients may choose. This explains why focusing exclusively on allowing and encouraging individual matching would fall into the same trap as many existing physician-centric proposals to resolve the conscience clause dilemma, namely, failure to deal with availability problems. Thus, Veatch urges the profession to adjust to create an adequate balance of physicians willing to satisfy a wide variety of patient demands.[35]

So long as the availability of a full range of medical services is preserved and all physicians bear the relatively minimal individual obligations explored in chapter 9, including the obligation to notify patients of their moral objections sufficiently far in advance that alternative pairings can be made if necessary, the higher-level contracts should reduce the transaction costs associated with permitting individualized contracts; not everything will be up for negotiation. For example, as Veatch explains, "there must be a presumption that clinicians are operating on some core, consensus morality unless they inform patients to the contrary.... [If a physician fails to so inform patients of his special ethical position], he must be bound either to accept the patient's terms or refer."[36] Because these inflexible baselines exist, it is not the case that allowing individual determinations of how a doctor's moral beliefs will be permitted to influence her medical practice will eliminate any semblance of overarching professional norms, as some critics of the matching model fear. In other words, it will still be meaningful to refer to oneself as a physician even if individual contracting is encouraged, although a contract/matching model may change the doctor-patient relationship into one more reminiscent of a relationship between a provider and consumer.[37] This is not inherently problematic, however, since such a transformation began long ago.

Presumably, Veatch is right that doctor-patient matching, with both doctor and patient probing the values of the other, will eliminate many crises of conscience, since it will involve notice and prior agreement about the scope of the relationship, rather than an attempt to negotiate responsibilities later on under the shadow of inevitable disagreement based on fundamental misunderstandings on the part of both parties as to what the relationship would entail. Morals matching may also have the added benefit of improving the level of trust that is so integral to successful medical encounters. A patient's trust in his physician allows the patient to let his guard down, despite the inherent vulnerability of the patient role. A trusting patient is able to share intimate details, feelings, and fears with the physician, and is also more likely to cooperate in the doctor's prescribed course of treatment, which in turn results in better care. For obvious reasons, the freedom to select one's doctor and care setting contributes to the development of this all-important trust,[38] and there are a variety of factors that influence a patient's selection, be it gender concordance, willingness to take weekend appointments, or a treatment philosophy that strives to avoid unnecessary medical interference.

Patient trust is likely built primarily on the belief that the doctor is technically proficient, but it is also bolstered by the interpersonal competence of the physician and indications that he or she is the patient's ally.[39] Interpersonal competence certainly relates to shared views on medical issues of special moral salience for the patient, and while a vision of the physician as patient ally generally refers to the notion that the physician will be willing to fight insurance companies and other barriers to gain access to important services for the patient, partnership between doctor and patient can be much deeper. Naturally, trust will be enhanced through alliance on moral issues. In fact, there has been a great deal of academic discussion about the value of incorporating spirituality and morals in the clinical encounter based on arguments that limiting physician communication to areas of professional medical expertise will result in neglect of other important aspects of the human condition.[40] The incorporation of patient spirituality could surely be facilitated by allowing patients to discuss these issues with like-minded doctors.

Focusing on benefits to the patient ought to render the matching model tolerable even to those who hold fast to the idea that a physician's personal preferences have no valid role to play in patient care, but some commentators would even limit the power of patient preference. They suggest that so long as patients can find physicians who will provide the services they want and will not force them to accept services (or even discussions about services) the patient finds morally objectionable, a sufficient match has been made. These critics argue that patients have no valid interest in the interactions of their physician with other patients and thus should be allowed to dictate only the rules of their personal interactions with their physician, rather than having a say over the physician's broader rules of engagement.[41] That is to say, if my gynecologist does not coerce me into having an abortion against my will, I have no legitimate objection if he or she provides an abortion to someone I sat next to in the waiting room. According to this argument, the profession would have no obligation to ensure that there are some physicians who simply will not provide the objectionable service to anyone if that is what some patients prefer. Instead, it would have a more constrained responsibility to ensure only that there are physicians who will provide the service to those who wish to have it. Notably, this version of matching may not offer much protection to would-be refusers because even if they were all excluded from the profession, patients could still allegedly find their physician match.

This argument, however, ignores the many benefits of moral diversity and conscientious objections described previously, and also fails to address the problem of moral complicity, which can be avoided through a more robust understanding of matching and its benefits. Patients do have a valid claim to avoid providing their support to medical services they find morally objectionable, even if that support involves no more than patronizing a physician who provides those services to others. Unsurprisingly, Pellegrino has reached a similar conclusion, arguing that forcing conscientious physicians out of the field entirely would be "unjust to patients who have a right to seek out and be treated by physicians who share their values; the choice of physicians should not be limited to so-called morally neutral physicians who do not."[42]

Perhaps more convincingly, a patient's legitimate interest in his physician's interactions with other patients is underscored by the importance of trust discussed above. Suppose a patient learned that his otherwise excellent physician had previously spent time in prison after being found guilty of a criminal offense such as child abuse, rape, or violent murder.[43] While such a physician may not even be permitted to return to practice, assume for the purpose of this hypothetical example that he was. Few would question that patient's unwillingness to continue seeking care from that physician, despite the fact that the source of the patient's distrust is completely external to the patient's relationship with the doctor. Why, then, would it be inappropriate for a patient who genuinely believes abortion to be the equivalent of murder to seek to avoid a physician who provides that service? If trust is central to the physician-patient relationship, and trust (or even belief in a physician's moral integrity) is undermined based on the services a physician provides to other patients, there is a strong argument for allowing a diversity of physicians—including refusers—in the profession to accommodate matching.

Like most arguments, however, even this trust argument can be taken too far since it could be used to justify allowing race, gender, and other potentially discriminatory factors to play a role in shaping the supply of physicians in order to satisfy consumer choice.[44] Accommodating some patient preferences may be unnecessary and even wrong—even if genuine trust issues remain for the anti-Semitic hospital patient assigned a Jewish doctor, for example, preferences based on racist, sexist, homophobic, or other sorts of bigoted views can be appropriately excluded from legitimate discourse as entirely illogical and not even arguably correct.[45] While it may be easiest to simply comply with a patient's discriminatory preferences and accommodation may improve that individual doctor-patient relationship, the broader harms of bigotry are too great. Placing some limits on patient preferences is justifiable, but some patient preferences—those that rest on at least an arguably rational foundation—are worthy of attention. Therefore, deep-value pairing is the ultimate goal, though critics are right to note that finding a physician who is willing to provide the desired service is a higher priority than finding one who is also willing to deny undesirable services to others.

Close-as-possible physician matches for *patients* are what really matter. However, it is important to recognize that there may not always be patient matches for *physicians*. Market demands will allow some moral beliefs to flourish and will force others to wilt. As we have already established, refusers should generally be permitted to enter and remain in the profession even if they are unwilling to satisfy certain patient requests, despite the fact that alternative measures of preserving patient access may be easier and cheaper to implement. Nevertheless, no one has any obligation to ensure that physicians are able to find a sufficient number of patients willing to accept limitations on the range of services the physician offers or to otherwise help keep refusers afloat. Refusers have no claim to the creation and preservation of an artificial market for their particular style of practice.

Of course, this is not at all to say that market demands necessarily establish what is morally correct and what is not. Free moral debate is important, and refusers can and should continue to espouse their arguments against the services they find objectionable in hopes of convincing others that their practice limitations are right. However, just like everyone else, they must find their own mechanism of support, whether it be from patients, academic positions, or some other source. Again, *room for refusal* and diverse moral perspectives is what is essential; artificially ensuring that all possible strands of argument are constantly represented is not. Even Mill did not advocate government assistance for unpopular views, instead limiting his opposition only to their obstruction and stifling.[46]

This leads us to a crucial point about the employment context, where the relevant question is not the broad issue of whether physicians should be permitted to exercise conscientious refusals as a general matter in their own practice, but rather is more specifically whether (or to what extent) employers should have to accommodate these beliefs. In other words, if a doctor refuses to perform abortions, must a Planned Parenthood clinic keep him on staff? The answer ought to be clearly no, and of course, a person with a genuinely held moral objection to abortion would probably never seek out or wish to retain such employment. However, if he did, the language of some existing statutes would leave the clinic helpless to terminate, transfer, refuse staff privileges to, demote, reassign, reduce the wages or benefits of, or otherwise penalize such a physician.[47] Unless a court is willing to creatively interpret such statutes in order to avoid these absurd results, as some in fact have, the clinic would have to retain the refusing physician *and* hire a replacement actually willing to do the job, doubling its costs.[48]

The shelter offered by some expansive conscience clauses is virtually limitless and lacks regard for the burdens imposed on the employer, going far beyond the protection for religious beliefs mandated by Title VII, or even the analogous accommodations required for certain categories of disabled persons under the Americans with Disabilities Act. Title VII actually permits employment discrimination on the grounds of bona fide occupational qualifications reasonably necessary to the normal operation of a particular business. It requires reasonable accommodation of an employee's (or prospective employee's) religious beliefs and practices only if such accommodations would not impose an undue hardship, defined as more than a de minimis economic cost, on the employer.[49] The Americans with Disabilities Act, while obviously not applicable to religious beliefs, similarly demands reasonable accommodations for employees or prospective employees, but limits its requirements to accommodations that would not require "significant difficulty or expense," and does not mandate the employment of an individual who could not perform the essential functions of the position even with reasonable accommodation.[50]

These restrictions on employer obligations appropriately recognize that the business of business is business, even as they try to avoid discrimination against vulnerable groups. Many conscience clause statutes, on the other hand, fail to allow for such commonsense limitations, though they should. As we saw earlier, conscience clauses are clearly doing something for physicians, but in this context, they are often doing too much.

Policymakers may argue about whether a de minimis threshold is acceptable or whether greater protection for physician conscience in the employment context is necessary, but some limits on the adjustments that will be required and caveats for bona fide occupational qualifications are essential. Employers should have no obligation to subsidize physicians' moral beliefs by hiring or retaining doctors whose conscientious refusals would require the employer to undertake significantly burdensome accommodations. If refusing physicians are simply allowed to enter and continue in the profession, are shielded from liability when they satisfy their limited responsibilities, and are reasonably accommodated by their employers, that is sufficient. Should they also find patients who wish to seek care from them despite or because of the services they refuse to provide on moral grounds and/or employers who wish to accommodate them to a greater extent, such as hospitals or clinics that share their moral beliefs and reservations, that would be ideal, but physicians need not be offered any greater protection from rejection by market forces.[51] Granted, this does eliminate a significant amount of protection for refusers, but not too much. It appropriately prevents refusers from shifting the costs of their views to their employers, while still allowing physicians, *who need not be anyone's employee*, to try their hand on the market and compete for patients against those with different conscientious limitations on their practices or none at all.[52]

Importantly, this seems to be a more desirable approach than not even giving physicians a chance to determine whether patients will find the matching model attractive. After all, "it is one thing for a true believer to try out her moral convictions in the public sphere and find them incapable of attracting sufficient interest and support to be viable; it is quite another for the state to forbid her from even trying."[53] When given a chance, however, it seems that the matching model is in fact capable of relative market success. Religiously sponsored hospitals have been in operation for years, with patients largely (though not always) aware of the fact that the services offered will be somewhat restricted, and morality-based managed care systems have also begun to develop around the country. The hospice system is another important, and widely accepted, illustration of doctor-patient matching, since a patient would not enter the program if her moral beliefs prohibited "giving up" on life, and a physician would not select palliative care as a specialty if his religion demanded that every medically possible option be explored to extend life regardless of quality.

Other instances of matching have also become increasingly popular. For example, the Tepeyac Family Center in Fairfax, Virginia, is an obstetrics and gynecology practice that is self-described as combining "the best of modern medicine with the healing presence of Jesus Christ," providing natural family planning services, but not other contraceptive devices, sterilization procedures, or abortions.[54] Similar specialty practices consist of a solo practitioner or a group practice made up of physicians with like values, while others operate with several physicians that hold various beliefs from whom patients can select internally. The clinics are careful to screen out prospective patients whose desires or needs are beyond their scope, which is unproblematic since physicians generally have no obligation to enter a professional relationship with every person seeking their services. A representative of the American College of Obstetricians and Gynecologists, Anita L. Nelson, seemingly approves—she explains that "if women know before selecting them, then it's quite a legitimate thing to do and might meet the needs of many women and doctors." Additionally, patients who go to these clinics appreciate the match they have found, expressing relief that they are understood by their physicians and are not judged for their views.[55] Here, matching appears to be a win-win situation.

A similar result is exemplified by the Pope Paul VI Institute for the Study of Human Reproduction, which attracts more than seven hundred new patients each year.[56] The system taught and used by the Institute "works cooperatively with the natural fertility cycle and enables doctors to treat women and married couples, especially Catholic married couples, in a way that allows them to live out their faith." Physicians report that this approach allows them to remain in practice without sacrificing their values, while patients also express satisfaction that they can "go about things in a morally sound manner." While these physicians have been labeled disingenuous[57] and these practices have been criticized as balkanizing medicine,[58] they appear to provide at least a workable element of the solution to the conflict of conscience problem since they avoid the creation of reliance interests. Patients who want more will at least theoretically be able to protect themselves by finding a willing provider in advance of their need.

Matching doctors and patients based on their deeply held moral values is quite promising and may not even be entirely novel. Aside from the examples just described, there are other circumstances in which patients already select physicians based on agreement with their personal beliefs, such as whether they take an aggressive or conservative approach to treatment, preferring the obstetrician who performs a battery of prenatal tests over one who does not or choosing the cardiologist who favors lifestyle changes rather than one who frequently recommends surgery. However, as we have seen, this tactic cannot suffice on its own to resolve the conscience clause debate, since match*ing* requires match*es*. Permitting unfettered moral subspecialization could inappropriately narrow the options available to patients, which would be particularly troublesome in poor or rural areas. Therefore, a pure contracting model is problematic at both a descriptive and normative level since patients may not have the knowledge necessary to make favorable agreements with their most desired providers. They may also lack the power within modern managed care systems to craft such free partnerships,[59] let alone within family structures in which a parent selects the physician for the adolescent patient, who may desire some services that the doctor refuses to provide. Additionally, when facing medical crisis, a patient may simply not have the time to seek a matching physician.[60]

Thus, the matching model must be accompanied by initiatives to secure the existence of reasonably accessible, diverse physicians, who, when considered collectively, are able to provide full access to the basket of medical services patients may desire. It must also be supplemented by mechanisms to ensure that patients can become rational contractors, which will require that they have sufficient information about the moral objections of their physicians at the earliest possible time and are aware of available alternatives. These constraints are imposed by the nonnegotiable obligations established by Veatch's primary social contracts, but matches will take time to accomplish, and in some cases, may not be possible at all. Nevertheless, a proposal in which matching is paired with systems to ensure availability, as well as a reiteration of baseline professional obligations, can have a significant impact in resolving the controversy surrounding conscience in medicine.

4

Which Institution? Licensing Boards Bearing the Burdens of Conscience and Access

So far, we have established that there is an important role for personal morality in the practice of medicine and that it would be far too harsh to expect the medical profession to satisfy its collective obligations to the public by excluding physicians who are opposed to the provision of particular medical services, at least without first exploring other options. Implementation of doctor-patient matching would be an important step in the right direction, but while matching is the ideal, we have not yet seen how the prerequisite of patient access can be satisfied. We begin down that road now, learning some important lessons from the sort of institutional solutions in place in both the pharmacy and end-of-life contexts that aim to preserve the autonomy of all interested individuals.

Patient access concerns are at the very heart of the conscience clause debate, with opponents properly arguing that physicians cannot be permitted to hold patients hostage to their personal moral beliefs.[1] The lack of harm posed to others by the exercise of a claim of conscience in the professional setting is a primary condition for the legitimate exercise of such a claim, but if patients can be assured of the reasonable availability of services they desire from some competent physician, they are not significantly harmed. Importantly, we must distinguish patient inconvenience from true lack of access, since only the latter warrants state intervention as a serious violation of the profession's duty.[2] Patients have no claim to have physicians at their beck and call, for if they did, the delays many patients experience in obtaining appointments, the hours spent in waiting rooms when physicians are running behind, and even the fact that patients must usually take time off from work to keep appointments would all be more problematic than they are. Of course, this is a normative claim, and some argue that any amount of inconvenience is an inappropriate burden on patients when that inconvenience stems from a professional's refusal to provide a service on grounds of personal conscience. However, we have already dismissed those views that refuse to give an inch to the opposing perspective and that have led the conscience clause debate into an apparent stalemate.

The more productive approach is not to ask whether allowing refusals by physicians would place any burden at all on patients who do not share the physician's beliefs, but rather to ask how much burden can appropriately be imposed. Many commentators predictably agree that refusal clauses protecting physicians that impose only minimal burdens on others are more acceptable than exemptions that demand substantial sacrifices from patients.[3] In fact, existing commentary coming closest to offering a compromise solution to the conflict-of-conscience problem seems to have accepted the distinction between inconvenience and lack of access, claiming that a physician who is morally opposed to a patient's request should be allowed to refuse, "so long as the patient is ensured safe, timely, and financially feasible alternative access to treatment."[4]

However, the "so long as" approach relies on ambiguous language and the passive voice for an easy escape from a difficult problem. The solution seems obvious when one says that the patient must simply be guaranteed access, but *who* exactly bears that responsibility? If it is the refusing physician herself, we have not gotten very far, since even if the individual physician has sufficient information to assess the presence of safe, timely, and feasible alternatives, which could impose significant costs on the refuser, she will have little leverage to change a scenario in which such alternatives are absent. Unable to evaluate and ensure the availability of alternate providers, the refusing physician will be forced to comply with the patient's request, which is precisely the problem we are seeking to avoid. An institutional body, on the other hand, will have easier access to information regarding the alternatives available to a patient, as well as the capacity to offer incentives and make other changes to improve the selection of physicians, thus creating more potential matches for patients and reducing the inconvenience posed by any individual physician's conscientious refusal.

Institutional solutions are often better than individual solutions in this context because they are capable of respecting the provider's interest in following the dictates of his own conscience, while the institution serves as the facilitator of all consciences, offering a substantial and workable compromise.[5] Further, an institutional solution avoids some inherent restrictions on the law's ability. Unless we are willing to accept heretofore unacceptably coercive measures and limits on freedom, laws cannot actually force individual physicians to provide services they morally oppose; if the physician genuinely objects to the service and the objection is serious enough, he will simply choose to face the consequences of refusal imposed by the law rather than comply with the patient's request. As we will see in more detail subsequently, this clearly would not resolve any access concerns for that patient, and, if the refusing physician is suspended, permanently disqualified from practice, or leaves the area to find a more amenable moral environment, it would also have the troublesome side effect of creating separate access concerns for patients requesting morally innocuous services.[6] These concerns are equally applicable to a system that acts prior to a physician's refusal, such as prohibiting licensure of those unwilling to provide all services within their specialty. An institution, however, could largely resolve these access problems by working to preserve a sufficient supply of willing providers.

Notably, institutional solutions, while sparking significant controversy of their own, have already demonstrated some relative success in other instances of moral conflict where there are readily available institutions on which to place the burden of ensuring patient access, namely, pharmacies and hospitals. For example, in the context of the highly publicized trend of pharmacists refusing to fill prescriptions for the morning-after pill, Illinois Governor Rod R. Blagojevich issued an emergency rule in April 2005, later finalized as a permanent regulation,[7] that requires pharmacies to accept and fill prescriptions for contraceptives that they carry without delay, but does not apply specifically to individual pharmacists. This initial attempt at compromise was not perfect, given that it prevented pharmacies from satisfying their obligation through referral, but was not accompanied by any limitation on the expansive conscience clause protection the state provides to employees.[8] However, a settlement agreement reached in October 2007 resolving several lawsuits challenging the regulation requires changes that will significantly mitigate this problem. Under the settlement, the state agreed to amend the rule to allow a pharmacy technician or store manager to fill a prescription in the face of a pharmacist's refusal, so long as the prescription is approved by an off-site pharmacist via phone or fax. This provides a more economical solution than simply having to staff an additional pharmacist to cover for a refuser.[9] Unfortunately, the modified regulation will still apply only to stocked contraceptives, allowing pharmacies conceivably to evade the regulation simply by not stocking those drugs at all.[10] Nonetheless, it does begin to address access issues without sacrificing individual conscience, and several states have followed suit.

The Washington Board of Pharmacy has similarly attempted to separate the obligations of pharmacists and pharmacies, but as we saw in chapter 1, its regulations have also been the subject of litigation. Under the challenged Washington rule, pharmacies have a "duty to deliver lawfully prescribed drugs or devices to patients and to distribute drugs and devices approved by the U.S. Food and Drug Administration for restricted distribution by pharmacies, or provide a therapeutically equivalent drug or device in a timely manner consistent with reasonable expectations for filling the prescription."[11] The rule does not require pharmacies to fill prescriptions containing an error or those that are contraindicated, fraudulent, or unavailable despite good-faith attempts to comply with stocking regulations, but outside of those circumstances, a pharmacy must fill a patient's prescription on site regardless of moral objection. The board has stressed that it is ultimately the pharmacy's responsibility to make sure that patients' prescriptions are filled, and pharmacies cannot comply with the rule simply by referring patients elsewhere.[12]

While the rule does not mandate that individual pharmacists dispense medications against their personal objections,[13] pharmacists challenging the regulations claim that is the effect of the burden on pharmacies. Indeed, some claim that they have been fired as a result of their refusals since their employers cannot afford to keep them on staff, despite the state's broad conscience clause protection. The court in *Stormans, Inc. v. Selecky* sought to avoid this effect, which it determined to likely be a free-exercise violation, by preliminarily enjoining disciplinary action against any pharmacy that refuses to dispense Plan B but immediately refers a patient to a nearby source, since this "refuse and refer" approach would not necessarily require that refusing pharmacists be fired in order to comply.[14] Regardless of the ultimate outcome of this litigation, which was still ongoing at the time this book went to press, note that this represents an additional option for preserving both patient access and the professional's conscience.

California has instituted a rule that gives pharmacies more leeway, since it permits a pharmacist to refuse fulfillment of a prescription only with the consent of his or her employer, but allows the employer to deny such permission if granting it would create an undue hardship. However, if permission is granted, the pharmacy must itself ensure timely access to all drugs, not just those used for contraception.[15] Delaware, New York, and Oregon have also taken steps in this direction, as their pharmacy boards have interpreted professional obligations to prohibit a pharmacist from obstructing or abandoning a patient and to require pharmacies to ensure timely access to prescriptions.[16]

Finally, movement at the federal level has been stalled for some time, but the Access to Legal Pharmaceuticals Act, introduced in both the House and Senate in April 2005, latches on to this institutional solution as well.[17] This proposed legislation would impose a burden on pharmacies, enforced through private causes of action and other possible penalties, to ensure access to all legal prescriptions for drugs used to treat health conditions for which the pharmacy normally keeps drugs in stock by having some pharmacist employed by the pharmacy fill them. Notably, individual pharmacists are not explicitly protected by the bill's language, and thus pharmacies could fire any refusing employee if they so chose, unless prohibited by the terms of some other statute or regulation, such as Title VII or state law employment protections.

As we can see, these examples each have problems of their own, which are often related to the fact that policymakers recognize that employers have the profitability of a business to maintain. Thus, the pharmacy-based approaches tend to either insufficiently protect access by limiting the pharmacy's obligation only to certain prescriptions or to insufficiently protect conscience by allowing the pharmacy to meet its more comprehensive obligations by excluding refusers. However, problems also arise from failure to respect the business needs of employers by placing conflicting responsibilities to patients and pharmacists on their shoulders. Of course, these concerns may simply indicate that the burden has not been placed on the right institution.

Nevertheless, we can begin to see how imposing the responsibility to ensure availability on some institutional body, rather than on individuals, could theoretically offer a compromise capable of protecting both patients and professionals by introducing a higher-level mediator. Certainly, patients may experience some inconvenience and potential discomfort from being denied fulfillment by any pharmacist, but these factors should be given minimal weight if the patient ultimately obtains timely access, particularly considering that none of the regulatory compromises described above tolerates the destruction of prescriptions, refusal to return unfilled prescriptions, or other forms of harassment by refusing pharmacists. More importantly, the patient may never even be aware of a pharmacist's refusal—if the patient drops off the prescription, any refusal could be handled internally without the patient ever knowing. Notably, in a public opinion poll conducted by Catholics for Free Choice, four out of five women supported policies allowing individual pharmacists to refuse to dispense contraceptives when the pharmacy bore the obligation of assigning another employee to fill the prescription; remove this safeguard and the same number of women opposed protection for refusers.[18]

Similar compromises have also been implemented to varying degrees by several courts facing conflicts of conscience between patients and providers with regard to end-of-life care. For example, New Jersey's intermediate appellate court decided *In re Requena*, a case involving a patient with Lou Gehrig's disease who wished to remain in the hospital where she had resided for several months while discontinuing her life-sustaining treatment.[19] This became problematic because such discontinuation was against the hospital's pro-life values, leading both the hospital and its individual personnel to refuse to participate in withholding artificial feeding. However, based on an assessment of the burdens on the patient that would be associated with transfer to a willing institution, the court required the hospital to abide by her request. The court recognized that its decision would also impose a significant burden on the hospital's nurses and technicians, but it justified this imposition on the grounds that as "well and whole people" with "full and vibrant lives ahead of them[, ... it is] fairer to ask them to give than it is to ask Beverly Requena to give."[20]

There are several notable aspects of this decision. First, the court's ruling was not based on any perceived professional obligations of the nurses and technicians to sacrifice their own interests to those of the patient, but rather was based on a sort of cost-benefit analysis—the professionals had life opportunities that were no longer available to the patient, so their sacrifice would necessarily be less. More important to the analysis of the institutional solution is the fact that while the court discussed the health-care providers who would have to comply with the patient's wishes, the burden it imposed fell squarely on the hospital itself, not on any particular individual. Conceivably, the hospital could have temporarily brought in additional staff not morally opposed to the patient's decision in order to avoid burdening its existing employees. Of course, this approach does not assuage any institutional claim of conscience that the hospital itself may have, but it does protect individuals, which is our focus here.

In a similar case, *In re Jobes*, the New Jersey Supreme Court required a nursing home to allow a patient to remain in residence despite the fact that the patient's family would not consent to continued artificial feeding, against the nursing home's claimed policy against discontinuing such life-sustaining treatment.[21] This court also recognized the burden its decision would impose on nursing home personnel, but determined that this burden was inferior to the "immense hardship" that would face the patient and her family if their wishes were refused. The court similarly discounted the institution's conscience but failed to impose the burden of removing the patient's feeding tube on any particular objecting medical professional.

As a final example, consider *Gray v. Romeo*, in which a Rhode Island court was willing to allow a hospital to arrange the transfer of a patient seeking to discontinue life-sustaining care against hospital policy, but cautioned that if successful transfer could not be promptly achieved, the objecting hospital would have to comply with the patient's wishes itself.[22] Again, the decision placed no specific burden on any individual objector, but rather directed its remedy at the institution. Thus, there is significant precedent for forcing institutions to provide services against their will, but not for forcing physicians to do the same. In fact, in *Grace Plaza of Great Neck, Inc. v. Elbaum*, the court held that even when a patient's conservator could not locate an alternative willing provider, "medical professionals should not be compelled by the State to engage in conduct which they themselves consider unethical."[23]

Note that while these courts facing institutional objections ultimately concluded that other interests were of greater consequence, none was necessarily denying the validity or importance of institutional conscience, and it is essential to recognize that institutions, like individuals, can truly harbor conscientious objections to various medical services. While some beg to differ,[24] Pellegrino explains that

organizational ethics is a systematic examination of the morality of collective actions in human institutions dedicated to some specific purposes in society. The ethical "code" or commitment of a specific institution is now customarily expressed in its mission statement. This is in a way the "conscience" of the institution. All who work in that institution are in some way accountable for adherence to the organizational mission, which is in effect a promise by the institution to behave in a particular way. Catholic institutions in America have for a long time had specific ethical directives that define them as Catholic hospitals. They are also committed to a charitable, even preferential treatment for the sick and the poor. Catholic hospitals can properly be considered to have a definable institutional "conscience," one, which given the content of the Catholic moral tradition, could and does come into conflict with secular society and its "values."[25]

Similarly, Lynn Wardle argues that health-care institutions, while not flesh and blood, are "organized by individuals to achieve purposes that can best be achieved by collective action."[26] To exclude them from protection, he argues, is merely an indirect way of denying protection to the individuals who comprise these institutions. At the very least, it seems clear that institutional conscience is something that should be taken seriously, and many legislatures treat it accordingly. Aside from the protection offered by the federal Church Amendment, forty-three states allow health-care institutions, not just individual providers, to refuse to provide abortion services,[27] nine states allow institutions to refuse to provide services related to contraception,[28] and fifteen states allow them to refuse to provide sterilization services.[29]

This seems to present a serious challenge to our approach, for perhaps an institutional solution is not a compromise at all. However, when an institution is ordered to perform an action contrary to its collective conscience, it may still be the case that no individual must *directly* breach his or her individual values; instead, the institution could find an outside performer. Of course, the problem of complicity with immoral action would nevertheless be present, since someone in the institution would have to call that outsider in order to make the request and sign her check, but no insider would actually have to perform the abortion, for example. This ability to distance oneself is not a perfect solution for many, and even these indirect violations are deeply troublesome to those affected by them. Thus, they should likely be avoided whenever possible, perhaps by allowing institutions to refuse services when there is an even higher institutional body that can mitigate access concerns, but the fact remains that distance to the objectionable act is morally relevant on a scale of acceptability.[30] Therefore, it is better for an institution like a hospital or pharmacy to have to violate its conscience, with individual members forced only to indirectly participate in behavior they find objectionable, than for individuals to have to directly perform the objectionable act. As between the institution and the individual, the individual should win out—our attempt at compromise remains intact.

For all these reasons, the institutional approach is most promising, but the problem that remains is defining the appropriate target institution capable of accommodating *all* physicians, bearing the burdens of access so that they do not have to. Unfortunately, institutional solutions seem to be more amenable to certain professions and specialties than others. For example, pharmacists and nurses tend to be both professionals and employees, and artificial nutrition and hydration often occur in the hospital or nursing home setting, so in conflict scenarios involving these individuals and/or services, there is an employer who can conceivably bear the responsibility of ensuring patient access. Even then, we have already seen that it may be asking too much of employers to simultaneously ensure access and accommodate refusing employees. This request becomes particularly onerous when we recall that according to the gatekeeper model of medical professionalism, it is the collective profession that bears the access obligation. Thus, we should be looking for an institution that can stand in as a tangible representation of the abstract idea of the profession as a whole.

Clearly, pharmacies and hospitals fail in that regard, though they may nevertheless work as a consequential matter because they can serve as useful mediators between professionals and their patients, in addition to the fact that employer-based systems are relatively simple to implement. These considerations may be a sufficient justification for employer-based compromises to conflicts of conscience, but regardless, physicians may not be institutionally employed at all, and often provide most of their care as solo practitioners or in localized partnerships with other physicians, even if they also have hospital privileges.[31] This is a significant problem that has been at least implicitly recognized by Alta Charo.

Of the existing commentators, Charo goes the furthest in her attempts to flesh out exactly how the collective obligations of various health professions may be satisfied. While she suggests that laws focusing on institutions rather than individual professionals may be appropriate, and even indicates that licensing laws that would permit professionals to join guilds willing to offer only a certain subset of legal services would be acceptable, her examples involve only pharmacies and religious hospitals. When discussing physicians, she notes only one way that their profession can fulfill its collective obligation—simply require every physician to provide all products and services under penalty of loss of license, loss of employment, liability to patients, or other professional discipline, eliminating any room for conscientious refusal. Charo recognizes that this is no compromise at all, and moves on to a discussion of the institutional solutions, but here she abandons physicians, offering no institution that could bear their collective burden for them.[32] Does her surrender suggest that the compromise is inapplicable to physicians or did she simply not delve deep enough? Is there any institution that can successfully facilitate the resolution of conflicts occurring outside the hospital and pharmacy, thus providing comprehensive protection to the individuals most personally involved in the provision of potentially objectionable medical services?

On initial consideration, we might think of the American Medical Association, or the various specialty societies, as the logical embodiment of the profession suitable to bear its responsibilities. However, while this may have been true in the past, when the AMA could claim as many as 75 percent of American physicians as members, today only one in three doctors belongs, and these numbers would be even lower if only those paying full dues were counted.[33] Further, membership in professional societies is purely voluntary; if the organization does anything a physician disagrees with or imposes any obligations he finds objectionable, the physician can withdraw his membership without consequence to his ability to practice. While licensing boards may rely on the principles established by these professional societies, giving them more weight, the organizations themselves simply lack the representativeness needed for legitimacy and the direct clout needed to influence physician behavior. Thus, neither the AMA nor any specialty medical society can appropriately serve as the institution responsible for ensuring patient access.

Another possibility to consider is insurers. Very few physicians outside of those who are highly specialized or provide purely elective services not generally covered by any insurance provider accept only out-of-pocket payments for medical care. Therefore, one might ask whether third-party payer networks could serve as appropriate institutions to bear the burden of ensuring access. After all, these networks already have the organizational capacity to measure supply and demand for physicians and various services. Whether they are willing to meet that demand is another story, but these companies do seem to have the necessary tools at their disposal. They might conceivably assess the expected need for all medical services in a given area and ensure that they have a proportionately sufficient number of physicians willing to provide those services. Then, if one physician refuses, the patient could find an alternative provider within a reasonable geographic radius. It is certainly possible that insurers would be prone to miscalculation, since they would not capture demand created by the uninsured, but even the uninsured might benefit from such a system, since a significant number of willing providers would have to be reasonably accessible, at least geographically.

However, placing this burden on private insurers is inappropriate and infeasible for a wide variety of reasons. First, insurance companies are not professionals and therefore lack fundamental professional obligations to provide needed medical services. Instead, while some are nonprofit organizations, many insurers are businesses that can, at least arguably, be appropriately guided by pure profit goals. More fundamentally, third-party payers cannot be said to represent the profession as a whole, even if they cover the vast majority of physicians. In fact, their interests are often at odds with one another.

Further, insurers would provide an extremely uncoordinated mechanism of ensuring the availability of medical services. If supply and demand for physician services did not adequately match as a result of physician refusals, which of the many insurers doing business in a particular area would bear responsibility? One might suggest holding the insurer covering the patient who was refused services responsible, but that solution would make the uninsured particularly vulnerable to physician refusal on grounds of conscience. There could also be tremendous free-riding and collective-action problems, with smaller insurers benefiting from the efforts of their larger counterparts to resolve access problems stemming from moral refusals or all insurers waiting for the others to take care of the problem first. Additionally, insurers have no obligation to cover all medical services and many do not cover abortion, contraception, or reproductive technology services, for example.[34] It would not make sense to expect them to ensure that supply meets demands for services they do not even cover, and it would certainly require an enormous policy shift to require insurers to cover all legal medical services.

Another major problem with naming insurers as the institution responsible for facilitating physician conscience and patient access is that in the face of such a burden, it would be far simpler for insurance companies to focus on access, neglecting physician conscience. Insurers would likely demand that all of their network physicians provide all covered services within their specialty regardless of moral objections rather than attempting to keep track of physician objections and ensure that there are enough willing providers. If the most common refusals relate to reproductive services, recording physician preferences and objections might not pose too great a difficulty, but once more controversial technologies become scientifically possible or are legalized, this task may prove far too cumbersome for insurers. For administrative convenience, the logical insurer would simply exclude conscientious refusers from its networks. In today's medical market, however, these refusers would then experience tremendous difficulty staying afloat and would essentially be forced out the profession or compelled to abandon their chosen specialty, thus diminishing the important pool of potential matches for patients. If placing the burden on an institutional body was intended to serve as a compromise of the interests of both doctors and patients, such exclusion would hardly achieve that function.

We should also note that health-care insurers might have institutional claims of conscience that we ought to attempt to protect if possible, though as we saw above, protecting individual conscience is more important. Even more persuasive is the idea that individual patient members of an insurance network might have moral claims themselves and may wish to avoid supporting the provision of medical services they find ethically objectionable, even indirectly through payment into a common fund.[35] While these considerations may not be enough on their own to disqualify insurers as the proper locus of compromise, combined with the other shortcomings just described, demanding that insurers bear the collective responsibilities of the medical profession appears to be a wholly inappropriate and likely ineffective attempt to balance conscience and access.

What other institution might be left to render the compromise solution workable for physicians? Fortunately, there is one other unorthodox proposal worthy of consideration before concluding that patient access concerns must trump individual physician claims of conscience—we might place the burden of ensuring access to an adequate supply of willing physicians on state licensing boards. Importantly, these boards are capable of acting as a proxy for the medical profession as a whole, and while they obviously consist of individual members, these individuals have not banded together to advance their personal views, separating them from other institutions that may have a valid claim to protection of conscience.[36] Licensing boards have mandatory leverage over all physicians, and they are present in all fifty states, the District of Columbia, and U.S. territories. Most importantly, unlike employers, they are responsible for establishing and maintaining the collective monopoly that at least partially creates the obligation of the profession to ensure the availability of medical services, since they selectively determine which individuals are fit to practice and punish those engaged in the unauthorized practice of medicine.

Licensing boards are in the position to mitigate the effects of that monopoly, protecting both conscientious refusers and patients, by taking measures to guarantee that the monopoly powers of the profession, and associated responsibilities, never (or at least rarely) trickle down to individual physicians—in other words, by ensuring a sufficient supply of accessible alternative gatekeepers and avoiding situations of the "last doctor in town." Goodin's principle of group responsibility suggests that when a person's interests are vulnerable to the actions and choices of a group of individuals, that group has a special duty to organize and implement a scheme for coordinated action such that the vulnerable person's interests will be protected as well as they can be.[37] Holding licensing boards responsible for ensuring that patient demands that would go unsatisfied in the face of conscientious refusals by individual physicians are nevertheless met satisfies the profession's group duty.[38] Under this approach, individual physicians could refuse to provide medical services to which they are morally opposed, subject to certain restrictions discussed subsequently, while a patient unable to access a willing gatekeeper could hold the licensing board itself accountable.

Recall from chapter 1 that but for existing conscience clause protection, licensing boards likely have the power to require that physicians provide certain services as a condition for licensure, regardless of their potential moral objections. However, blanket requirements that completely ignore individual variances in the moral beliefs of both doctors and their patients, and thus ignore the value of moral diversity in medicine explored previously, are precisely what a compromise solution hopes to avoid. If a compromise is to work, licensing boards—as government agencies that have responsibilities above and beyond those of business owners for the welfare of all members of society—must not shirk their institutional obligation by placing it right back on the shoulders of individual physicians, even if that would be cheaper and less complex. Instead, they must balance access against respect for conscientious refusers. Note that this approach does not require licensing boards to create an artificial market for refusers, but rather simply allows them to offer their subspecialty to the public while patients who prefer different services are simultaneously accommodated.

At least one other commentator has recognized the value of a compromise reached by imposing a "governmental responsibility to ensure supply of and patients' access to [nonobjecting] practitioners."[39] Bernard Dickens has pointed out that "it often lies with governmental licensing authorities, health professional associations and healthcare institutions such as hospitals and clinics to ensure that licensed healthcare professionals who object to participation in delivery of lawful services can and do refer patients, in a timely way, to other professions who do not object."[40] Dickens relies on a human rights analysis as the basis for a government obligation to simultaneously respect physician conscience and ensure that patients retain reasonable access to legal medical services from nonobjecting providers.[41] Unfortunately, however, he has failed to elaborate on these suggestions, providing no guidance as to how any government body could go about ensuring availability, what level of availability should count as acceptable, and what to do when things go wrong. Thus, Dickens offers an important step in the right direction, but does not take us far enough toward the implementation of a novel and promising compromise capable of accommodating the interests of both doctors and patients. This book, on the other hand, attempts to offer as many concrete details as possible.

III

The Details of the Institutional Solution

Measuring Patient Demand and Determining Which Demands to Meet

Because conflicts of conscience in medicine can be alternatively framed as a function of nothing more than mismatched supply and demand for physician services, one solution is to make sure that there is a sufficient supply of competent, willing physicians so that patients desiring a particular medical service can realistically obtain it.[1] If this task can be accomplished, we have seen that there is no reason to demand that individual physicians deliver care in conflict with their conscience, and in fact, that there may be several reasons to allow physicians to refuse. However, as is so often the case, pinpointing the desired result is far easier than actually attaining it. There is an incredible array of definitional concerns and measurement questions that must be resolved at both an empirical and normative level before we can say that supply and demand are inappropriately imbalanced, that any such discrepancy should be fixed, and that any intervention aimed at doing so has been successful, or at the very least, sufficient.

Resolving unacceptable inconsistencies between supply and demand resulting from the conscientious refusals of some subset of physicians will first require licensing boards to measure these factors, as well as to determine appropriate geographic areas in which their measurements of supply and demand will be meaningful. If the problem is truly one of access, licensing boards will also have to establish just what is meant by that term, much less "timely" or "prompt" access. What is the dividing line between lack of access and inconvenience? Is it enough to ensure that the patient can find an alternative provider in the same office building as the refuser? Within the same traveling time for the patient? In the same city, county, or state? Further, should the medical necessity of the patient's request bear on the way feasible access is defined? If so, who has the authority to evaluate necessity—the patient, the physician, or someone else? Boards must also assess whether they have a responsibility to satisfy all sorts of patient demands or whether there are some requests that physicians are appropriately refusing on grounds of conscience, as well as whether there are just some grounds for refusal that are unacceptable, even in the absence of related access concerns. Finally, once each of these questions is answered, boards will have to take action—upon identification of patient access problems stemming from physician refusal, boards must work to resolve them, but how exactly should they go about doing that, and what should happen if they fail?

Clearly, state licensing boards have their work cut out for them, but the challenges they face are far from insurmountable. The remainder of this book explores each of these issues in depth in order to fill in the contours of the institutional compromise, providing the most complete guidance possible to boards charged with avoiding the very access problems that are chiefly responsible for opposition to conscientious refusals in health care.

Assessing patient demand for various medical services provided by physicians should be relatively simple as compared to the other tasks facing licensing boards, especially given that insurance companies and hospitals must routinely collect such data for their planning purposes. State governments, too, already compile similar information, since they must ensure adequate availability of services to Medicaid/Medicare beneficiaries. Boards could begin with these figures, and academic health services researchers may also play an important role, given that they would be free to study demand for things like elective cosmetic surgeries performed in the physician's office and would not be constrained to those services covered by health insurers or provided in hospitals. Licensing boards may even consider establishing grant funding or creating private, nonprofit organizations devoted to obtaining the necessary data.

Potential financing mechanisms will be discussed subsequently, but for now, suffice it to say that needs-based planning is one existing, although imperfect, way for experts to advise the government, insurers, and other interested parties on the per capita number of physicians necessary for the adequate management of medical conditions cared for by a given specialty.[2] Current utilization rates of various physician services may provide an additional source of helpful information, although care must be taken not to perpetuate rates that are artificially low as a result of an existing insufficient supply of physicians willing to meet patient demand.[3]

Although there are some significant difficulties associated with accurately quantifying need or demand for health services, the important fact is that licensing boards would not have to start from scratch in developing measurement techniques. Researchers have crafted a variety of methods by which to forecast health human resources supply and demand, including techniques borrowed from demography, epidemiology, economics, and industrial engineering, which could be adapted by boards interested in the institutional compromise to resolve the conscience clause dilemma.[4] Reasonable estimates of patient demand are feasible, and are all that could be expected of licensing boards.

However, a more pressing dilemma remains—precisely which demands for medical services must licensing boards measure? In other words, for which services should these boards bear a responsibility to ensure access to willing physicians despite conscientious refusals? Does their responsibility cover all legal medical services, is there some room for discretion based on prevailing views of medical ethics, or is some other dividing line more appropriate? These are extraordinarily difficult questions because many seemingly appropriate dividing lines would be either under- or over-inclusive. They ultimately exclude socially important services like abortion that must be included in order to make this compromise even remotely palatable to those opposed to conscientious refusal by physicians, or include things like the provision of futile care that probably should be excluded given consideration of what most commentators currently view as appropriate grounds for denial of access.

Before moving on to evaluate a variety of potential standards, however, it is essential to note that the limits, if any, adopted for a board's responsibility in this context have *no bearing whatsoever* on what individual physicians can or cannot do. If a board determines that it must ensure access to sex-selective abortion, for example, individual doctors could still permissibly refuse to provide that service. Alternatively, if sex-selective abortion does not fall within the licensing board's responsibilities, this does not mean that individual physicians would be prohibited from providing it, unless it were otherwise declared illegal. Further, under the Constitution, boards would have wide latitude to exclude services from coverage. While the state would not be facilitating access, we saw previously that there is no general constitutional right to the provision of medical care; failure by the state to actively ensure the availability of services would not be an undue burden even on reproductive freedoms, particularly when unaccompanied by any restriction on the services individual physicians could choose to provide.[5]

Constitutional analysis, of course, does not necessarily resolve ethical issues and the institutional solution to conflicts of conscience will not succeed if boards hide behind constitutional minimums. Thus, while licensing boards have no constitutional obligation to guarantee access to physicians willing to provide any medical service, they do have access obligations stemming from other sources. Given the profession's collective monopoly power and the public's dependence on it for all sorts of legal medical services, combined with the profession's institutional representation by state licensing boards, any limit on the board's obligation to satisfy patient demands will have to survive rigorous examination.

That being said, the public has routinely expressed moral disgust and outrage, although certainly not uniformly or unanimously, at some of the things that physicians have been willing to do. Thus, perhaps some limits truly are in order, and it is likely that not all access concerns stemming from conscientious refusals are worthy of attention or resolution. For example, serious objections have been lodged against such things as the "Ashley treatment," where a severely disabled child was intentionally kept from undergoing puberty in order to make it possible for her parents to continue caring for her,[6] as well as the selective implantation of embryos with the gene for dwarfism or deafness,[7] the provision of growth hormones to children who are not sick, but are just short,[8] gender reassignment surgery, cloning, and all of the situations introduced previously to demonstrate the value of physician restraint.

Proponents of moral limits on the services the profession must make available go on to offer more extreme examples such as the case of Dennis Avner, who goes by his American Indian name, Stalking Cat, and is known around the world as the "Catman" due to his numerous surgeries and cosmetic procedures intended to make him appear more like the animal.[9] Given the theory that this patient's demands are the result of a form of body dysmorphic disorder, refusal to satisfy them may be permissible on the basis of medical expertise; in fact, concerned commentators argue that Avner's doctors should have long ago begun denying his requests, claiming that he is being harmed rather than helped by medicine. However, if there is no other available treatment to mitigate the patient's disorder, this may come down to a values question—one that cannot be answered with certainty either way. The procedures he seeks are certainly legal, but should licensing boards be responsible for making them available?

For a somewhat less bizarre example, consider a 2007 *Today Show* interview with a sixty-year-old woman who had just delivered healthy twins.[10] The mother defended her decision to have children at this point in her life by stating how powerful it made her feel, explaining that this was just another example of the freedom of modern women, and asserting that she hoped to be a role model for others. When asked directly about the worry that just because we can do something does not mean that we should, she responded that "people need to get ready for what's coming up in our society.... There are a lot of middle-aged women [having babies]—forties, fifties, now I just turned sixty. That's going to be acceptable. They have to just keep up with what's going on with society." Unfortunately, she failed to seriously address concerns about her life span and the impact it could have on her new children, though she did express a sense of security as a result of the fact that her partner was significantly younger and would likely be able to care for the children as she aged.

The physician present for the interview had not been involved in implanting the embryos, which occurred in a South African center specializing in in vitro fertilization of older women, and while he did caution that having children late in life is risky for both the mother and child, he refused to condemn this woman's actions. Instead, he explained: "I can't be judgmental about that. This has taught me to be very open-minded." While stories of mothers the age of grandmothers have become more prevalent and are often dubbed miracles,[11] large segments of society do not seem willing to accept that any woman at any age should be able to become a mother, and many of the online responses to this particular story denounced the woman's choice as selfish and inappropriate.[12] Major health concerns aside, which are clearly significant, commentators wonder whether it is morally acceptable to assist in the creation of a child who is very likely to lose at least one parent at a young age.

Similarly, there is a growing outcry over the fertility industry's continued practices leading to the birth (and often death) of high-order multiples, such as poorly monitored use of fertility drugs and the implantation of too many embryos at once. As with older mothers, these pregnancies are frequently referred to as blessings, particularly since the users of assisted reproductive technologies have often struggled for so long to have children. Unfortunately, however, they pose a tremendous health risk since the human body is not designed to carry so many fetuses at once, leading to maternal complications and extreme prematurity, which in turn can result in enormous financial and emotional costs.[13]

After watching patients who chose not to abort one or more fetuses to heighten the chances of saving the others walk away with severely disabled children, or none at all, several physicians have decided that they can no longer work with patients who will not agree from the start to selective reduction of a multiple pregnancy as an option. They generally recognize that women have the right to be against abortion, but point out that they did not enter the field of assisted reproduction to help create tragically damaged babies; those women will have to find other doctors who are able to accept such terrible outcomes.[14] Some doctors refuse to implant more than a single embryo at a time,[15] but even they recognize that patients who have invested a great deal in having a baby are often unwilling to bank on such a conservative tactic. We've waited long enough, the patients argue, why wait through several cycles and pregnancies if we can have two (or more) babies at once? They are our embryos and it is our choice, they say. So they shop around until they find a doctor to meet their demands, and often they are successful, particularly since many clinics are eager to boost their statistics for helping women achieve pregnancy and implanting several embryos does increase the odds.[16]

But what if patients demanding to have four, five, six, or more embryos implanted at once could not find a doctor willing to do that—is that a problem boards ought to be expected to resolve? Even if individual physicians can refuse, should *someone* have to help Catman, senior citizens looking to become biological mothers, or other patients resting on claims of personal autonomy to justify doing things that are legal but morally questionable? These examples suggest the need for an appropriate limiting principle with regard to the services for which boards ought to bear access obligations in their endeavor to resolve conflicts of conscience. However, establishing such a reasoned standard turns out to be incredibly tricky, for it is quite difficult to find a dividing line that we can legitimately claim would be accepted by the public when negotiating and agreeing to a professional monopoly over the provision of medical services.

Perhaps, then, we are left with the idea that in order to avoid the illegitimate imposition of the moral views of some onto patients who disagree in areas where we really cannot be sure who is right and who is wrong, licensing boards should have a broad responsibility to ensure access to a combination set of physicians willing to offer all legal medical services. In fact, some commentators seem to believe that this is the case, although it is unclear whether they have fully considered the consequences of their claims. For example, we saw in the last chapter that Bernard Dickens has argued that health-care institutions have an obligation to help patients seeking "lawful" services make their way to some professional who will not object.[17] Similarly, Robert Veatch has pointed out that forcing individual physicians to violate their conscience should be rare because there will often be another physician available who is "willing to provide a *legal* service that patients believe to be ethically necessary."[18]

Given that these discussions do not explicitly state that legality alone should be the sole criterion for either individual or institutional obligations, there is reason to believe that both Dickens and Veatch would be willing to consider legality simply as an important starting point, but that additional limiting factors would be relevant as well. Otherwise, the results might be absurd. Licensing boards could avoid having to ensure access to doctors willing to provide late-term abortions, assist with

patient suicides and perform female circumcision where illegal, and prescribe drugs and devices that have not yet been approved for any use by
the FDA. On the other hand, boards would have to satisfy patient demand for essentially *everything* else, including medical services that simply do not work, so long as they have not been prohibited outright by
the legislature. Thus, under this bare standard, any herbal remedy or
off-label use that a patient desires would have to be some physician's
command.

The benefit of a simple legality standard is that it avoids all moral
judgments that have not already been made by the legislature and thus
offers patients significant freedoms. Patients would only be constrained
by the views of others that have found success through the democratic
process. However, it feels as though something is missing here, and
what that something is can be determined by once again considering the
social negotiation that resulted in the profession's monopoly. Part of
what the public expects from the medical profession is technical proficiency, but it also relies on physicians for resolution of the many difficulties posed by information asymmetry. Doctors have scientific knowledge
as a result of their extensive training that patients involved in all sorts of
other endeavors simply lack the time, and perhaps the capacity, to build
for themselves. Physicians know which interventions will achieve the
desired results and which will not, and patients depend on them for this
expertise.

Thus, it seems appropriate to add to the legal limit so that licensing
boards would have no obligation to ensure access to a medical service
unless, as a scientific empirical matter, that service has a reasonable likelihood of success in achieving the desired result. Under this standard,
licensing boards could avoid having to satisfy demand for treatments
that are unproven or have been disproved and need not ensure access to
physicians willing to provide interventions widely regarded in the scientific medical community as pure quackery, such as the use of vitamin C
injections or laetrile to treat advanced cancers. Even this addition, however, would provide only a very weak limitation on a licensing board's
duty because what counts as a desired result must be left up to the patient as a subjective matter; it is clearly beyond what can be determined
by medical expertise within the exclusive realm of physicians alone.

Consider, for example, the Jehovah's Witness patient who refuses surgery involving blood transfusions, consenting only to bloodless surgery. Empirically speaking, certain risks of bloodless surgery are in fact greater, to the point that some physicians refuse to provide it, claiming that they cannot in good conscience and in accordance with their professional oaths place the patient in such danger. Ignoring the potential inconsistencies in such claims, especially considering that the risk of no surgery at all is often greater than the risks associated with bloodless surgery and those are the only two options the patient is willing to consider, an individual physician may appropriately refuse to perform bloodless surgery on any patient.

However, the question here is whether the licensing board representing the profession as a whole can correctly refuse to satisfy patient demand for this service when its goal is to resolve access problems associated with conflicts of conscience. The scientific data shows only that it is risky, but it does not—and cannot—show that the risk is too much for a patient to bear. Of course, the issue may be more complicated for the risks associated with some of the reproductive technologies introduced above, since in those situations patients are not only accepting risks for themselves, but also for their children. However, the fact remains that determining levels of acceptable risk is not a matter of purely objective judgment.

Similarly, evidence demonstrates that natural family planning methods, such as those involving calendar charting of the menstrual cycle and monitoring of cervical mucus, hormones, and body temperature, are more likely to result in unintended pregnancy than other contraceptive methods.[19] However, this is not necessarily because NFP is a scientific sham, but rather because many women use outdated techniques, have difficulty implementing them, or fail to use the methods correctly 100 percent of the time. Nonetheless, more than half a million American women rely on natural markers of fertility to achieve or avoid pregnancy,[20] choosing to tolerate the risks and failures in order to comply with the tenets of their religion or because they deem the risks preferable to those associated with more intrusive methods. Their assessment cannot be labeled wrong as a matter of fact and it would likely be inappropriate for licensing boards to simply ignore demand for medical services that facilitate the sort of personal choices that ought to be a patient's to make.

Ensuring the availability of these services may not be very problematic from a normative perspective, but bigger problems linger just around the corner. Because choices about acceptable levels of risk are not a matter of medical expertise, but rather are a matter of goals and values, the addition of the scientific limitation would not permit licensing boards to set standards defining futile care or allocating scarce medical resources. Instead, they would be forced to ensure access to at least arguably inappropriate patient services—certainly a controversial conclusion.

Futility is an extraordinarily difficult medical concept because, while the determination is often made by physicians, in many circumstances, it is truly a value-based decision that the benefits of additional medical interventions are not worth their costs. There is no dispute that neither physicians nor licensing boards must provide access to scientifically invalid treatments, treatments with no medical indication for the patient, or treatments that offer no possible benefit.[21] However, unless the medical service will have absolutely no physiological impact on the patient or technically cannot accomplish the patient's goals, whether the service is capable of producing a worthwhile outcome is a determination that should be made by the patient, or potentially by society given the costs of such services and the scarcity of resources, including a physician's time and attention. This determination is just not amenable to a scientific conclusion,[22] and even those states that permit hospitals to discontinue the provision of futile care of their own accord only allow the futility determination to be made by an ethics committee in consultation with the patient's family.[23] In much the same way, there is nothing unique about the profession that renders it especially capable of assessing the considerations of equity and distributive justice inherent in allocation decisions.

Perhaps limiting otherwise expansive licensing board obligations only by legality constraints and conclusions of scientific impropriety regarding the medical services in question correctly reflects the social negotiation resulting in the professional monopoly held by physicians. Without a doubt, there are some who would grant patients the extensive right to demand that the medical profession provide them with anything and everything they take to be beneficial. However, those who remain troubled by the breadth of licensing board responsibilities the legality and scientific expertise standards would impose, particularly those who believe that just because medicine can do something, either technologically or legally, does not mean that it should, will continue to search for additional limiting principles in hopes of avoiding the sorts of examples described above. But could any further limitations on a board's access obligations be justified?

If boards may consider further factors in hopes of avoiding only inappropriate denials of access resulting from conscientious refusals, John Stuart Mill's harm principle might be useful.[24] By this standard, licensing boards would be required to meet any patient demand so long as that demand would not result in harm to others and is not the result of deficits in the patient's knowledge or rationality. In this latter case, however, the proper solution is not to forcibly prevent the patient from going forward, since, as Mill explains, only the patient himself is capable of judging "the sufficiency of the motive which may prompt him to incur the risk." Instead, we may only warn him of the danger in order to remedy his deficit of understanding.[25] Should his demand remain thereafter, and would harm no one but himself, licensing boards could not properly ignore it, perhaps obligating them to ensure access to things like female circumcision and assisted suicide where those services are not legally prohibited.[26] Note that this result highlights the first of many potential difficulties with application of the liberal approach to the context of board obligations, since it would require government assistance with self-harm, as opposed to mere noninterference.

That issue aside, the harm-to-others rationale for limiting accommodation of patient requests appears to be exemplified by the response of some Canadian physicians to the technological feasibility of prenatal sex selection. Canada has no law against sex-selective abortion or against determining the sex of a fetus before birth. Nonetheless, the College of Physicians and Surgeons of British Columbia, a statutory body created by the Provincial Legislature and charged with protecting the public through the regulation of medical practice, created a policy not to disclose the sex of a baby until after twenty-four weeks gestation in order to reduce the risk of gender-selective abortion, which it considers morally inappropriate. The basis of this determination was concern that facilitating sex selection would result in several negative externalities, including the reinforcement of discriminatory attitudes based on sex, violation of the principle of equity between males and females, and starting down a slippery slope toward other eugenic decisions that are socially repugnant.[27]

However, reliance on these broad social harms to restrict individual liberty is troublesome from Mill's perspective. Because we are part of a society, our actions can rarely avoid having some rippling effect that impacts others, but extending the harm principle to all such nebulous effects is quite dangerous. On that understanding, there would be no violation of liberty the principle could not justify.[28] Thus, under a broad version of the harm principle, licensing boards might be permitted to disregard too many patient demands on the basis that they are likely to cause harm to others. On the other hand, if harms must accrue directly to defined others, potentially real harms that are more circuitous and less discrete, such as those prompting the Canadian policy, may be inappropriately ignored. Therefore, we run into the opposite problem, which is that a strict application of Mill's philosophy might not serve as much of an additional limit on the legality/scientific expertise standard at all, since it could require boards to ensure access to doctors willing to provide a wide range of services unless some more specific type of harm to others stemming from those services could be identified.

Unfortunately, in addition to this trouble regarding which type of harm ought to suffice to eliminate a service from the scope of the board's obligations, the presence of harm itself is not always an objective, empirical matter—it is not always as simple as pointing to the number of patients who opted for assisted suicide following the legalization of that practice because they were coerced, clinically depressed, or could not afford proper care. Instead, it can be quite malleable and subject to reasonable disagreement. For example, considering they would not otherwise have existed, have premature quintuplets conceived with the help of fertility drugs who are now facing serious disabilities truly been harmed?

Further, as we saw previously, abortion at least arguably imposes a harm directly on others, certainly for those who believe the fetus is a full human person, and likely still for those unwilling to go so far. Some commentators have also argued that abortion imposes even more widespread negative externalities that range beyond the fetus by establishing a culture without sufficient respect for life.[29] Thus, licensing boards applying either the narrow or broad version of Mill's harm principle plausibly could—and in the current political climate, many likely would—refuse to undertake the responsibility to make abortion services available regardless of the conscientious objections of individual physicians. However, what about the fact that some people reasonably argue that there is no harm at all? This sort of disagreement over what is at root a metaphysical question is precisely why it may be appropriate to allow individual physicians to refuse. But for licensing boards to do the same would have the very silencing result we are seeking to avoid, since access to certain services may be completely eliminated without recourse when objections are made at the institutional level.

That the harm principle could allow licensing boards to limit their obligations in this way is problematic on other grounds as well. First, this example demonstrates the principle's capacity to disproportionately affect women given their unique ability to bear—and thus to directly harm—children. Few medical procedures not related to reproduction can affect others in such a concrete and specific way, potentially leaving demand for them among those boards must satisfy, unless the broader conception of social harm is used, which is itself quite troublesome.

Additionally, even those who are willing to concede that abortion entails a direct and concrete harm to another of the sort envisioned by Mill may convincingly argue that the harms associated with eliminating access to legal abortion (or failing to ensure access) are greater. This weighing of harms is an issue that would likely give rise to even more controversy and disagreement. There is a strong argument, however, which we will consider shortly, that at least having the option of pursing a safe abortion is essential to the health, safety, and equality of women. Therefore, it is likely a component of the profession's—and the licensing board's—obligation to society, and we ought to be extremely wary of a principle that could be used to avoid it.

Finally, from a practical standpoint, excluding abortion from the licensing board's list of responsibilities would essentially eliminate the compromise nature (and goal) of the solution to the conscience clause dilemma proposed in this book, since any solution that fails to ensure access to abortion will be roundly rejected by one side of the debate. For all these reasons, the harm principle is likely to cause more problems than it will be able to solve and is thus not an advisable supplement to the legality/scientific expertise standard.

An alternative standard for separating services that licensing boards have an obligation to ensure access to from those they do not regardless of individual physician refusals could be based on whether the services in question represent genuine patient "needs" or only mere "preferences." Norman Daniels has argued that in terms of health-care, needs are all that we owe each other as members of society,[30] and some health-care workers raising conscientious objections have put forth this distinction as justification for their refusals to satisfy patient requests. They claim that they have no obligation to provide a medical service to "customers," but rather only an obligation to treat sick "patients," problematically attempting to distinguish between medical care and non-medical care that happens to use medical services. These health-care workers are perfectly willing to tend the wounds of a vicious criminal, but they consider birth control, abortion, assisted reproductive technology services, and the like to be mere "lifestyle choices, not treatments for diseases," and thus beyond the scope of their obligation.[31]

However, while one goal of medicine has historically been caring for the sick, certainly the profession also has an obligation to people who are not sick per se but are seeking medical services to preserve or enhance their health and/or overall well-being. Additionally, if the goal is to avoid licensing boards having to satisfy potentially immoral patient demands, the needs/preferences distinction will be an unsatisfactory proxy, since many mere preferences are nevertheless widely considered morally permissible and many things that could appropriately be classified as needs, perhaps such as future life-saving treatments derived from embryonic stem cells, are morally controversial.

Further, the distinction between needs and preferences itself poses significant problems, with definitions running the gamut from very narrow to very broad. For example, Daniels describes health needs as those things we require in order to "to maintain, restore, or provide functional equivalents (where possible) to normal species functioning (relativized to the appropriate reference class by gender and age),"[32] but this approach could inappropriately limit board obligations, as we will see below. On the other end of the spectrum are very broad understandings that would offer hardly any limit at all and that seem to beg the question, such as those classifying as needs any goods that are essential for the good of the patient[33] or that would at least provide a significant health benefit.[34] In fact, the term need in the health-care context often acts as nothing more than a placeholder for cost-benefit judgments.[35] Thus, we tend to say that you need an appendectomy because the cost is relatively low compared with the benefit of continued life, but that you do not need Botox injections, since the cost of that intervention is relatively high compared with the marginal benefit of temporarily having fewer wrinkles.

Rather than discrete concepts, needs and preferences appear to be plotted along a continuum, which leads to the obvious problem of trying to determine precisely whether and when a patient's strong preference ever becomes strong enough to resemble a need. Is the woman who seeks the morning-after pill to avoid pregnancy following a rape expressing a need or a just preference? What about the patient who wishes to terminate the life-sustaining care that is keeping him alive in agony and with no hope of improvement, or the more extreme example of an amputee "wannabe," a person suffering from apotemnophilia, who claims that he cannot escape from his obsession and contribute as a socially productive citizen unless a perfectly healthy part of his body is amputated?[36]

All of these things appear to be essential to the patient's welfare, broadly defined, and thus may be part of the board's responsibility, but whether they are required in order to preserve or restore species-typical functioning is a more difficult question. The amputee wannabe is trying to attain a sense of mental normalcy, which suggests that his claim may in fact be a need, but satisfying it would also destroy physical normalcy, making it unclear what the licensing board's access obligation would be. Terminating life-sustaining care would result in death, which in some circumstances is species-typical and in others is not; this leads us to a particularly troublesome twist, which is that under the needs/preferences approach, licensing boards might conceivably be able to ignore all demands for care coming from patients who have met or exceeded the average life expectancy. Finally, the example of trying to avoid an unwanted pregnancy may lead to a similarly disconcerting result.

Because a normal pregnancy is certainly not a disease, even when unwanted, and in fact may be the epitome of species-typical functioning, licensing boards might be able to avoid ensuring the availability of physicians willing to perform nontherapeutic abortions, and perhaps willing to offer other birth control methods that are requested for contraceptive purposes rather than to avoid disease.[37] However, depicting the ability to control one's reproductive capacity as a mere lifestyle choice ignores the fact that access to precoital contraceptives, and potentially to the morning-after pill and abortion services, is in many ways essential to a woman's dignity and freedom from the control of others—things that ought to be understood as far more than simple preferences. Many physicians consider these services to be an important part of basic health care, allowing women to avoid unwanted or high-risk pregnancies, manage reproductive health problems, and space and restrict the number of children they decide to bear. In turn, each of these things will permit them to take advantage of educational and economic opportunities, rendering them better able to improve their lives and the lives of their families.[38]

However, even if all licensing boards could be convinced that abortion services truly are needs and are thus part of their responsibility, the greatest problem with this needs/preferences standard is its inability to address the discrepancy that arises due to the fact that the medical profession has collective monopoly power that does not map onto any understanding of the needs/preferences distinction. A patient may simply prefer to have liposuction rather than to adopt changes to her diet and exercise regime, but the fact that she does not really need cosmetic surgery does not change the fact that if she wants it, she can only get it from a licensed physician. These services are not available on a wholly free market, at least not under circumstances that patients should be asked to face, and that barrier was not imposed to prevent patients from gaining access entirely, at least not in this case. Instead, it is meant to protect patients from quack providers. Thus, the exclusion of other potentially willing providers strikes the fatal blow to claims that the profession has no obligation to ensure patient access to mere "lifestyle" services.[39] If those services are to be excluded from the licensing board's mandate, it will have to be on other grounds.

Unfortunately, other potential dividing lines that might be utilized by licensing boards to limit their access obligations are similarly inappropriate and incapable of excluding only those services that ought to be treated as beyond the profession's mandate and no more. Distinguishing responsibilities based on whether the patient is seeking an emergency or nonemergency service would be problematic, since this would severely narrow the obligations of the medical profession, completely ignoring patients with chronic diseases and those seeking preventative care. It may also prove difficult to reach agreement on which patient demands are true emergencies—should the term be limited to life-and-death situations or is the morning-after pill also a medical emergency, given the limited window in which it must be taken in order to be effective? Allowing licensing boards to ignore demand for controversial medical services would not work either, though it has been suggested as an acceptable ground for conscientious refusal,[40] since it could subject patients to the whims of vocal minorities opposed to a medical practice and is not a good proxy for whether the service is ethically appropriate.

Finally, one may wonder why licensing boards would bear a responsibility for ensuring the availability of services that are not covered by major insurers, Medicaid, or Medicare. While this list might provide a useful jumping-off point for licensing boards, especially since it will be difficult to draw physicians to an area to provide services that patients cannot pay for, it is important not to confuse "is" with "ought." Just because various services are not covered under these programs does not mean that they are appropriately excluded or are unethical. Remember that the licensing board is representing the profession and must strive to satisfy the profession's broad obligations based on its socially negotiated collective monopoly; businesses and government assistance programs have an entirely different mandate.

Alas, none of these potential additions to the legality/scientific expertise standard are tenable, but must we really impose on licensing boards charged with resolving the problems of conscientious refusal an obligation to ensure access to doctors willing to implant deaf embryos, transform men into cats, amputate healthy limbs, or help octogenarians to conceive? Even if individual physicians could still refuse to provide each of these services, serious moral qualms linger. Part of the concern about eliminating a moral role for licensing boards may stem from a fear that simply because the legislature has not actively prohibited a given medical service does not necessarily mean that the service is morally acceptable. There are many widely recognized barriers to passing legislation, even when it is quite popular or seemingly objectively right;[41] just because something is not made illegal by positive law "does not establish societal indifference.... Law, because it governs many, must deal with generalities, leaving specific application to the good judgment of those charged with its enforcement. It leaves some conduct unregulated, not from a desire to encourage that conduct, but from the prudential judgment that the harms arising from declaring the conduct illegal are too grave."[42] Thus, there may in fact be a legal vacuum in which unethical or inappropriate services are nevertheless sliding under the radar, especially at the wild frontiers of medical technology.

Further, just as Chief Justice Taft recognized that Congress must be allowed to delegate lawmaking power to appropriately constrained administrative agencies if it is to exercise some of its powers at all,[43] it would be unreasonable to expect legislatures to review every medical procedure as technology advances and provide either a yea or nay. In such a climate, it may be appropriate to view licensing boards as state agencies delegated not only the power to regulate the medical profession, but also the moral authority to regulate medicine itself, satisfying patient demand for some services but not others. Such a delegation would be far more efficient than forcing this task on the legislature,[44] though the legislature would have to provide at least some guiding principles. Here, when applied at the institutional level, Savulescu's argument that it is inappropriate to deny services that are legal, beneficial, part of a just health-care system, and efficient may prove useful.

However, the definition of what is beneficial, just, or efficient should not rely entirely on a wholesale adoption of the profession's ethics in the absence of broader social input. This is because, as we saw briefly above, the profession has a claim to scientific medical expertise, but no inherent claim to moral expertise.[45] As Savulescu has appropriately noted, doctors do possess greater knowledge of the effects of medical treatment, but to say that this knowledge simultaneously suggests that doctors should collectively decide which treatments are appropriate for patients is nothing more than paternalism.[46] It is certainly true that individual doctors may have more experience with ethical dilemmas and have thus thought about them more deeply than the average person, but there is nothing special about doctors' training that makes them generally wise with regard to these matters.[47] In fact, "doctor and non-doctor alike start from the same point in developing their understanding of medical morality."[48] Further, even if physicians were appropriately given ethical deference in the past, as American society becomes more and more pluralistic, people will predictably come to support different trade-offs and will rely on different values, rendering individual physicians less able to predict or understand the values of any particular patient, and thus less competent judges of the "best" or "right" choice.[49]

If there is any such thing as moral expertise, which is in itself debatable given that no one has special access to knowledge about what is right and wrong,[50] individual physicians qua physicians have no special claim to it. Therefore, while there is significant value in maintaining a role for physician morality in the broad social debate, the opinions and beliefs of individual doctors demand no special weight in that debate, nor does that necessarily change when they stand as a collective, despite the fact that professional groups often take positions on controversial moral issues. It is for this reason that the claim that a physician can legitimately refuse a patient's request if most other physicians would also refuse that request fails. John K. Davis explains that this argument is not based on the assertion that majority consensus determines moral truths. Instead, it relies on the fact that the refusing physician would not be putting the patient in any worse position than he would have been had he sought care from a different physician in the first place, since that alternative physician would also have been likely to refuse.[51] Importantly, however, Davis fails to account for the fact that both physicians might be unjustified in their refusals even if they were allegedly relying on the internal morality of the medical profession, particularly if they failed to consult other stakeholders.

The court in *Brophy v. New England Sinai Hospital, Inc.* made the same mistake when it allowed a hospital and its medical staff to refuse

to participate in withdrawing life-sustaining treatment at least partly because the moral basis for such refusal was "recognized and accepted
within a significant segment of the medical profession and the hospital
community."[52] Despite the intuitive appeal of the court's rationale, neither physicians nor the profession of medicine as a whole have inherent
standing as experts on human values. If questions of this sort cannot be
appropriately left to individual patients, they are matters of serious concern for society at large.[53]

Thus, in order for any delegation of ethical discretion to the profession
via licensing boards to be legitimate and justifiable, licensing boards
must only be permitted to avoid satisfaction of patient demands they
would otherwise have an obligation to meet under the legality/scientific
expertise standard *if* the exclusion is based on the receipt and deep consideration of input from interested stakeholders. These include not only
doctors, but also patients, scientists, payers, ethicists, and other members
of the public. Broad social discussion could avoid the problem of simply
replicating inappropriate decisions by individual physicians on a wider
scale,[54] and a diversity of decision-makers is important if one believes,
as many commentators do, that medical ethics are far too weighty to be
left in the hands of physicians alone.[55]

As Veatch has explained, unilateral professional determinations as to
which services are morally appropriate handed down by an oligarchy of
physicians are wholly invalid, akin to a college fraternity claiming that it
is the "ultimate source of norms for the brothers because it invented the
rules."[56] This problem becomes particularly evident when one realizes
that most medical codes of ethics suggest that patient interests are the
foremost concern, but simultaneously allow physicians to refuse to participate in certain types of care based on personal moral or religious
objections. Clearly, without external moderation, the profession's norms
are too likely to reflect its own self-interests to be depended on to protect
those of patients.

This is not to say, however, that we can entirely discount the profession's ethics, for they have quite likely developed as part of the negotiation of power with the public, and are thus at least somewhat reflective
of more than the insular perspective of doctors themselves. Swartz argues
that generally accepted professional ethics are appropriate grounds for

refusal of care by individual physicians because these standards "have been developed by consensus and, over time, these standards can easily be shared in advance with the public."[57] Importantly, however, even within a single profession or specialty, many of the issues raising conflicts of conscience are not amenable to consensus and are often left to the discretion of individuals by professional codes. For example, the AMA takes no stance on abortion, and when the president of the American College of Obstetricians and Gynecologists appealed to U.S. senators to require physicians opposed to abortion to at least refer patients elsewhere, he was met with serious opposition from the American Association of Pro-Life Obstetricians and Gynecologists, a special interest group within his own organization.[58]

Particularly with regard to these issues where consensus is unlikely, the profession's ethics are not likely to offer significant guidance to licensing boards. Instead, if licensing boards are to have moral discretion with regard to the services that must be available to patients beyond the standards of legality and scientific propriety, they should rely on a fair process with serious deliberation and inclusion of a wide variety of stakeholders in order to vet all sorts of significant reasons why a particular medical service should or should not fall under the licensing board's responsibility.[59]

Daniels has argued that when we have constraints that provide general guidance about obligations, but that fall short of being able to actually resolve disputes and we lack consensus about other principles, we can proceed to set limits through the use of a fair process with accountability on the part of the body that makes the ultimate decision. In this context, our general constraints thus far are only that the licensing board cannot satisfy demands for services that are illegal and need not ensure access to services that are scientifically incapable of accomplishing the patient's goal. These serve as upper limits on the licensing board's potential obligations, but legislatures likely ought to also establish lower limits describing the services for which boards *must* satisfy patient demand, even if stakeholders make arguments to the contrary. This is crucial, since without such lower limits, licensing boards' obligations could be restricted by mere political controversy, a dividing line rejected above and one that would eliminate much of the value of the institutional solution.

Note that states already have some experience identifying which services fall under a reasonable understanding of a comprehensive healthcare scheme given their review and approval of health insurance products, often mandating the specific coverage of even politically controversial services. They could do something similar in this context, outright listing the services that boards must ensure reasonable access to or simply offering a more general definition of the lower limit below which boards cannot fall. While others may work as well, one possible baseline might be that licensing boards are responsible for at least ensuring access to those services that have attained the status of standard of care. This would entail services that a physician in a particular specialty would be reasonably expected to offer patients in the absence of conscientious objection, including abortion services, for example, even if some physicians would nevertheless be unwilling to provide them.[60]

Within the upper and lower limits, whatever they are decided to be, there are still likely to be some reasonable disputes. Implanting only embryos that will develop into deaf or dwarf children is legal and works to accomplish the patient's desired result, but it has not yet become standard of care. In that case, for example, licensing boards could work to ensure access to willing providers, but would not necessarily be required to do so. Should they? The same questions arise with regard to care that is sometimes described as "futile" from some perspectives but that is considered worthwhile from others. These value judgments that our general principles do not resolve and about which reasonable people may disagree are precisely where fair process comes into play.[61]

Licensing boards could convene diverse working groups to determine their obligations, with membership and procedural requirements similar to those established for Institutional Review Boards charged with reviewing research proposals involving human subjects. These groups could make use of existing recommendations or policies promulgated by voluntary professional societies as a starting point for deliberation. Before reaching final decisions, however, boards should take advantage of the type of fair process that is already in effect in the context of administrative law—notice-and-comment rulemaking. In this way, rather than relying on the decisions of only a subset of interested stakeholders, all interested parties could weigh in.

To satisfy their mandate and enforce the authority delegated to them by Congress, federal agencies promulgate regulations or "rules." Under the Administrative Procedure Act,[62] rulemaking may be initiated as a result of a public petition, internal agency decisions, or legislative intervention, but the key point is that the agency's rules cannot be a result of entirely internal deliberation. To drastically simplify things, it is important for our purposes only to note that the agency must first publish notice of its intention to adopt a rule dealing with a particular topic and must allow some reasonable period during which any member of the public can submit comments on the proposed agency action. Before promulgating a final rule, the agency must consider the comments it has received and must provide a concise and general statement of reasons for its ultimate decision consisting of an account of the agency's decisional factors, a discussion of plausible alternatives to the route the agency chose, and a response to each of the major issues raised by the public comments. To pass judicial scrutiny, the agency's rule must not be arbitrary or capricious and must reflect a genuinely, as opposed to hypothetically, rational decision.[63]

Similarly, licensing boards, as agencies, might propose the exclusion of particular medical services (e.g., those that are not standard of care, but are legal and scientifically valid for the patient's desired purpose, though nevertheless potentially morally questionable) from the ambit of the board's obligation to ensure patient access to willing providers. Any interested member of the public could offer arguments in favor of or against the board's proposed decision and the board could decide accordingly. This avoids the problems associated with aggregative notions of democracy under which people are compelled to abide by a majority rule not necessarily because it is right, but rather simply because that rule was derived from aggregating the preferences of the voters.[64] This aggregation prevents the resolution of fundamental moral differences through the appropriate mechanism of argument and deliberation. Notice-and-comment rulemaking, on the other hand, focuses on the persuasiveness of the arguments offered by public comments, not on the sheer number of similar comments submitted.

Additionally, notice-and-comment rulemaking has the benefit of being able to place constraints on the kinds of reasons that can play a role in deliberation, which Daniels suggests is a crucial requirement of fair process. He notes that when fundamental values are at stake, "people should not be expected to accept binding terms of cooperation that rest on types of reasons that they cannot view as acceptable," such as reasons that rest on religious faith. Notably, exclusion of these concerns is not inherently problematic in the context of the conscience clause debate since individual physicians will be able to rely on their religious values in their personal refusal decisions. Remember that here, we are focusing exclusively on the obligations of licensing boards to satisfy patient demand. Thus, licensing boards can limit their justifications to those that will be accepted as at least relevant by all, so that even those members of the public who disagree with the board's final decision can accept that the decision was based on reasons that appropriately played a role in the deliberation.[65] They may remain unconvinced, but the board's decision will certainly be more legitimate than denial of access to a particular service based solely on the moral beliefs of individual physicians without wider debate.

Finally, licensing boards can be held accountable for their actions in a variety of ways. First, they are at least in theory amenable to democratic influence if the public disagrees with their decisions. Should they improperly include or exclude a service, the legislature can intervene, either by legally prohibiting that service or by explicitly stating that the service falls under the board's responsibility. They would also be subject to judicial review if they ignore public comments or otherwise behave unreasonably. Both of these safeguards work to satisfy Daniels's revisability and appeals conditions of fair process.[66] Thus, for those unwilling to accept that licensing boards have an obligation to ensure access to all that has not been legally prohibited or proven to be scientifically invalid, but who recognize that ethical judgments can only be appropriately made by licensing boards that solicit input from those outside the medical profession, the notice-and-comment model offers a workable solution. It does not provide fine-grained standards according to which boards should make their decisions, but it offers the next best thing: a fair process for decision making with the opportunity to revisit those decisions that should have come out differently.

Before moving on, however, there is likely one additional limitation that ought to be placed on boards' obligations to ensure access. This limitation has nothing to do with morality, legality, or the scientific validity of a given service, but rather has solely to do with the practicality of the licensing board solution, and is necessary to avoid the "problem of one" that could arise if boards are held responsible for satisfying even the most idiosyncratic patient demands. For example, it is likely safe to assume that the highly unique requests of Catman and apotemnophiliacs could be excluded through the notice-and-comment process, since they certainly do not fall within the standard of care and are therefore not part of the board's baseline responsibilities. If they are not excluded there, boards would have to ensure access, although we will see subsequently that the relevant geographic boundaries in which access must be ensured could simply be drawn much larger for services that are in low demand. Even then, however, it may simply be too much to expect boards to devote their institutional resources to this sort of individualized request. Instead, legislatures may find it more appropriate to confine board obligations to the satisfaction of patient demands made at least by some reasonable proportion of the public, so long as this caveat is not used to exclude services that must be made available in any minimally decent health-care system.

If one of the major problems created by allowing individual physicians to refuse to perform various medical services on grounds of conscience is the elimination of access to services that the profession has a responsibility to make available, licensing boards will need to take action to make sure that access is not inappropriately eliminated. To do this, they must measure patient demand for a variety of medical services so that they can determine how many willing physicians are needed to satisfy that demand. However, there is a normative question raised by the fact that perhaps the refusing physicians are right—perhaps there are some services that patients truly should not have access to and for which the profession really should not open the gate. If that is the case, licensing boards need not ensure access to willing physicians, but what dividing lines can be used to make these assessments?

We have seen that many potential standards for differentiating between the patient demands licensing boards must satisfy and those they can avoid responsibility for tend to present results that do not seem to be in line with the primary reasons the public accepts the professional monopoly on the provision of medical services, namely, assurance of technical proficiency and mitigation of information asymmetry.[67] Thus, it may be appropriate to require licensing boards to satisfy patient demands for all services that are legal and empirically capable of achieving the patient's desired result. Under this approach, boards would have to ensure access to physicians willing, among other things, to perform abortions, assist patients with natural family planning, withdraw artificial nutrition and hydration, provide assisted suicide where legal, selectively implant embryos with the genes for deafness, dwarfism, or other things traditionally considered disabling, offer bloodless surgery, or even provide services that are hugely expensive and offer an extraordinarily low chance of extending a patient's life for a significant period of time.

However, those who are unconvinced that patient autonomy is the ultimate good and who believe that the legality and scientific expertise standards allow some unethical or otherwise inappropriate services to continue when they truly ought to be avoided will advocate more constrained responsibilities for licensing boards. Because the medical profession has no special claim to moral authority or expertise, however, licensing boards cannot simply refuse to satisfy patient demands on grounds of insular professional ethics, although they might permissibly refuse on the basis of broad public input, essentially serving as a legislative gap-filler. Under this delegation model, licensing boards that adopt a fair process of decision making can legitimately restrict their responsibilities in a way that allows them to play a role in avoiding the immoral use of medical technology without excluding services that are widely recognized as part of a comprehensive health-care system even if politically controversial.

Some patients will inevitably have to deal with the views and beliefs of others being imposed on them, preventing them from obtaining services they may want—but such is the nature of a cooperative society. This will be the case under the legality and scientific expertise standard, and even more so under the delegation model, but in both cases, such an imposition will occur less than under a system where individual physicians refuse their services and nobody has an obligation to ensure access to things that should remain available. State legislatures should determine for themselves the types of limitations on patient autonomy they find appropriate, amending their medical practice acts to provide licensing boards with guidance as to which standards they should use when determining which patient demands they, as the representative of the medical profession, have an obligation to meet. For sure, some patients will be silenced, just like the doctors who are unable to find enough patients willing to deal with their moral constraints to stay afloat, but under this model, everyone at least gets the chance to be heard, which is certainly more than can be said for the current state of affairs.

6

Measuring Physician Supply and Limiting the Grounds for Physician Refusal

Once licensing boards have determined which patient demands they have an obligation to meet and the level of demand for those medical services, their next step toward ensuring sufficient access despite conscientious refusals will be to measure the supply of physicians willing to provide those services. As was the case with measuring patient demand, measuring physician supply should prove relatively simple as a technical matter, with normative concerns getting the brunt of our attention instead. Here we consider which limitations on supply—which grounds for conscientious refusal—boards should accept as legitimate.

Because refusing physicians are opting to narrow their portion of the profession's obligation and because it would simply be most efficient, they should be required to register their conscientious refusal status with licensing boards on initial licensure and whenever their moral positions change. Boards would thus have clear data about the availability of various medical services falling under the profession's obligation, at least as affected by physician conscience, since they could assume that licensees who register no objections are morally willing to provide all services covered by their specialty. If refusing physicians failed to register these objections (perhaps through a simple check-box form), they might be appropriately confronted with a choice between compliance with the patient's request despite moral objections or forfeiture of conscience clause protection, resulting in some sort of disciplinary and/or liability consequence for refusal. These consequences could theoretically vary depending on whether the patient was able to reasonably access an alternative physician, although it is not clear that the penalty facing the unregistered refuser should depend on such fortuity.

Notably, this system of voluntary registration in order to exercise protected conscientious refusal is already in place for California pharmacists, who must notify their employers in writing of the drugs they object to dispensing before they can refuse to dispense those drugs.[1] A somewhat similar registration requirement has also been suggested by Judith Daar to deal with physicians' refusals to provide care they feel to be futile in the hospital context. Daar proposes that hospitals establish Treatment Evaluation Boards that would be "charged with adopting procedures for the appointment of an alternative physician when the treating physician requests the patient be transferred from his or her care."[2] She suggests that these boards could require all members of the medical staff to complete a questionnaire designed to elicit information about their professional conscience, specifically tracking their area of practice, years of experience, and interest in being placed on a list of doctors willing to accept patients transferred from other physicians as a result of futility concerns.

Texas has adopted an analogous approach to deal with disagreements between providers and patients regarding whether the life-sustaining care they are receiving or demanding is appropriate. In a relatively unique move away from the sort of concession to patient autonomy that has become so prevalent, Texas allows physicians and health-care facilities to challenge a patient's decision to proceed with or refuse life-sustaining treatment, and potentially allows care to be terminated against a patient's wishes. Resolution of such challenges is attempted by an ethics committee, but state legislation requires that in the meantime, patients also be given a list of health-care providers and referral groups that have expressed their readiness to consider accepting a transfer of patients seeking to either continue or withdraw care, or to assist with such a transfer.[3] This registry is maintained by the Texas Health Care Information Council and includes the identity and contact information of willing providers and referral groups located both within and beyond the state.[4] This system is the converse of the one proposed here, in that it involves the voluntary registration of those willing to perform some act rather than the registration of refusers, and is also far more limited than the type of refusal registry that state licensing boards would need to maintain given its focus only on life-sustaining care. Nonetheless, the fact that one state has developed such a list is a sign of the idea's achievability.

In addition to offering a feasible, inexpensive, and simple way of measuring the supply of willing physicians, a registration system also has the important benefit of creating some mechanism to evaluate the propriety of a physician's refusal in a timely fashion, for as we will see in this chapter, not all grounds for physician refusal ought to be accepted—conscience is not quite a "magic" word. If physicians were permitted to simply refuse on the spot when the patient requests a service that the physician opposes, valuable time might be lost while the patient is forced to seek review of the physician's refusal by the licensing board. Clearly, some medical services would not be amenable to this sort of forced waiting period. In an ex ante registration system, on the other hand, licensing boards could conceivably test the validity of a physician's reasons for refusal, as well as the sincerity of his beliefs, prior to any actual conflict. Physicians whose reasons are deemed unacceptable or insincere would then have the status of unregistered (and therefore unprotected) refusers, with its attendant consequences, whatever those may be.

Assessing the Validity of Moral Refusals

Because individuals hold so many idiosyncratic moral beliefs, and because conscience is so personal, it is quite difficult to assess the validity of a physician's moral objections. There is also the serious problem of identifying who gets to make the ultimate determination as to which reasons are valid and which are not. For example, one court has held that personal economic beliefs that are publicly and freely announced are a perfectly acceptable reason for a physician to refuse his or her services to particular patients—in this case, Medicaid recipients who would not consent to undergo a tubal ligation following delivery of their third child—but clearly not everyone would agree with that conclusion.[5]

Others have argued that in order to be acceptable, a physician's reason for refusal must be so integral to his or her moral framework that denying the refusal would represent a greater harm than that resulting from the restriction of patient access accompanying the physician's refusal of services.[6] While weighing harms is important, this standard is question-begging since it offers little guidance as to how harms to various parties should be weighed. Worse, it ignores the fact that denying protection for physician refusals may be harmful not only to that individual physician, but also to society more broadly. Further, if restriction of access is all that matters, the institutional solution will resolve that problem and all types of physician refusals could be deemed valid, but this conclusion is inappropriate, for other harms to patients may also be worthy of consideration. A more specific and selective standard of validity is needed.

Importantly, however, validity in this context cannot refer to the actual logic or truth of the moral beliefs in question, since requiring refusers to demonstrate such truth would most often be asking for the impossible. No one can empirically prove that abortion kills a person with full moral status, for example, or that contraception is an offense against God. The metaphysical nature of these objections and the inability to determine whether or not they are right is precisely why preserving room for conscientious refusal is so important. Adding a requirement that the beliefs themselves be factually accurate before those beliefs will be protected in the professional sphere would render any such protection of physician conscience effectively meaningless.

Additionally, there is no compelling reason to treat religious and secular moral objections to a particular medical service differently. As Kent Greenawalt has noted, those with religious convictions that a medical procedure violates God's law seem to have the strongest justification for declining to participate, but it simply does not seem fair to exempt only religious medical practitioners and not others with similar feelings and attitudes,[7] especially considering that religious beliefs represent only a single species of human motivators. Other very powerful guides for appropriate behavior exist in the lives of both the religious and nonreligious alike.[8] However, whether they are religious or secular, eccentric moral stances that are completely outside any reasonable understanding of a good physician's role—such as refusal to provide pain-management services on the grounds that a patient experiencing pain deserves it as punishment for poor character or needs to suffer in order to be accepted into heaven—should not be protected for any reason.[9] In fact, the "question of whether to allow conscientious objection may well turn on whether the ethical position of the doctor ... connects intelligibly with the core values of medicine."[10]

Of course, states could constitutionally choose to treat religious objections differently.[11] For example, the language of the Universal Military Training and Service Act, which governs the military draft, technically protects only objections based on religious beliefs, rather than those based on "essentially political, sociological, or philosophical views, or a merely personal moral code."[12] However, even in that context, the definition of religious beliefs has been construed quite broadly to include seemingly secular beliefs that hold a place equivalent to religion in the objector's own scheme of things.[13] One may be tempted to favor religious objections simply because they tend to be easier to prove—the objector need only point to concrete scripture or consistent statements from religious leaders. However, this ease of proof is no reason to reject as a matter of course moral objections that may be just as genuine, important to one's personal integrity, and potentially right. Thus, licensing boards can, and should, allow protection for secular moral objections as well.

So while validity cannot be measured by truth and should not be determined on the basis of where one's moral objection comes from, the matter cannot be simply set aside, for it is clear that even within the realm of what is appropriately called conscience, we should not necessarily provide a blank check to all refusers. Some physician objections may be grounded in suitable reasons, but others may in fact be highly intolerable. For example, a physician could (1) refuse to provide or withdraw a *certain service* across the board because she is sincerely morally opposed to the provision or withdrawal of that service in all cases for all patients, such as nontherapeutic abortions or cessation of life-sustaining treatment, or (2) refuse to provide care to *certain patients*, either because the physician genuinely believes that the requested service is (a) inappropriate for that *particular* patient, such as care the physician deems futile or too risky, but could be appropriate for other patients in other circumstances, or (b) inappropriate for that entire *group* to which the patient belongs, such as discrimination against homosexuals or single women seeking reproductive services, without concern for individual circumstances.

For reasons that will be explored below, the interests of doctors, patients, and society at large would be best served if licensing boards accepted as valid refusals based on ground (1) above, consistent opposition to the service for any patient, and rejected invidiously discriminatory refusals, ground (2)(b), recognizing that this dividing line may be at times either over- or under-inclusive. However, with regard to patient-specific refusals that are not invidiously discriminatory per se, ground (2)(a), the answer is not as clear. There, the refusal is based on patient characteristics rather than on the service itself, but the characteristics at issue may arguably justify disparate treatment, such as age, risk factors, or responsiveness to the intervention, because they are relevant in ways that factors like race and sexual orientation are not.

Note that there is some overlap between these proposed categories. For example, a physician who refuses breast implants to a particular fifteen-year-old because he believes that young girls are too vulnerable to peer pressure and have not yet had a chance to develop a stable body image, but provides them to a fifty-year-old, is in fact objecting to an entire group of patients, namely adolescents. Similarly, a physician who refuses to offer invasive and costly surgery to a terminally ill patient likely to die within weeks with or without the intervention, but who would provide the surgery to an otherwise healthy patient, is discriminating against the terminally ill.[14] However, these distinctions between groups are far different than a decision to refuse implants only to African-American teenagers or to refuse the surgery only to homosexuals, particularly since they are not inherently meant to judge or punish the patients to whom services have been refused, but rather are intended to avoid some other problem that just so happens to be associated with the patient's personal characteristics. The point is simply that the use to which the service is being put may be relevant in a way that is truly not bigoted, and licensing boards should have the discretion to make that call.

Thus, when the grounds for a physician's patient-based objections are not odious or inherently offensive, and are at least arguably relevant to the physician's role—even if truly a value judgment beyond the physician's strictly medical expertise—his refusal should be granted protection as valid. This conclusion is particularly appropriate in light of the fact that the patient could likely obtain the service elsewhere unless the grounds for refusal were so strong that they convinced the licensing board as well. To decide otherwise would wipe away a tremendous amount of discretion physicians currently enjoy and would relegate them to the role of mere technician forced to simply follow patient orders, which is precisely what this proposal tries to avoid.

Permissible Objections to the Service

For many of the same reasons, licensing boards should also accept a physician's registration of refusal based on his genuine conscientious objection to a service itself, an objection that stands regardless of who is requesting that service. Not only might the service truly be morally wrong, but importantly, patients should suffer little to no dignitary harm as a result of licensing boards permitting sincere service-based refusals. One is not harmed by the bare fact of having another person disagree with him or her, and nothing more is occurring in these cases so long as the reasonable availability of medical services is preserved.

Objecting to the service is not always equivalent to objecting to the patient. In fact, in this context, the only thing that refusers object to about the patient is his or her choice to pursue the service in question. Physicians who believe bloodless surgery is just too risky and refuse to provide it are not necessarily condemning the beliefs of Jehovah's Witness patients, they are simply not adopting them, just as physicians who oppose abortion need not judge those who feel it is their only acceptable option under the circumstances. Even when they do engage in judgment, however, they are denouncing the patient's choices, and not the patient per se. In a pluralistic society such as ours, it is clearly quite commonplace for people to have drastically different opinions of moral acceptability. If simple unwillingness to "embrace the values of another constituted harm, then all of us would be harming countless others all the time."[15] While refusing physicians must certainly refrain from browbeating or outspoken condemnation of patients seeking their services, if physicians make use of the registration system described above, patients will be able to assess a physician's moral beliefs before ever seeking her services, and can thus preemptively avoid any chance of dignitary harm.

This preconflict doctor-patient matching may be particularly important given evidence that some patients are more than just offended, but in
fact experience humiliation, whether justified or not, when denied legal
medical services or transferred elsewhere to obtain them.[16] For example, news stories abound of women describing a sense of embarrassment
as a result of being denied birth control and morning-after pills at the
pharmacy counter or by their doctors.[17] In Germany, a policy of mandated referral by health-care institutions with moral objections to a
patient's request for reproductive services was considered so incredibly
degrading that it was met with strident opposition—apparently, nothing
short of actual provision of the service when and where the patient
demanded it was considered sufficient.[18] However, even assuming for
the sake of argument that there are some actual dignitary harms stemming from allowing physicians to refuse under the limited circumstances
explored here, despite the fact that there are reasonable alternatives
available to the patient, those harms are not sufficiently bad to warrant
forcing a physician to violate his personal moral beliefs. Notably, any
other conclusion would eliminate room for compromise.

More troublesome, however, is the concern that permitting physician
objections to particular services would permit discrimination against patient groups who could not be discriminated against directly. Though
physician refusals are increasingly applicable to a variety of services
desired by both sexes, Alta Charo points out that for the moment, "the
focus of most refusals has been on actions associated with sexual or reproductive choices of women."[19] By this logic, these refusals are more
aptly described as backhanded objections to the patient than objections
to the service; those who refuse to provide abortions or prescription contraceptives are technically treating men and women equally, but only
women are being denied access to services that have been deemed an essential component of basic health care. For precisely this reason, others
have argued that the refusal to dispense prescription contraceptives in a
pharmacy is a form of sex discrimination that violates public accommodations laws, just as employment discrimination cases have held that failure to provide insurance coverage for contraceptives in an otherwise
comprehensive health plan is impermissible.[20]

These analogies do not map out perfectly onto the physician context, however, since unlike pharmacists or insurers, physicians often specialize in the treatment of a single gender. Thus, ob-gyns who will not perform abortions are not in fact discriminating against women as a function of their unique ability to become pregnant, but are instead truly objecting to a particular service, just like urologists who will not perform vasectomies. Nevertheless, the collective impact of that sort of refusal may end up having a problematic discriminatory effect, for if all ob-gyns refuse to provide abortions, women would be made worse off as compared to their male counterparts.

While actions resulting in a disparate impact on a certain class of persons are not unethical or illegally discriminatory in every circumstance, Charo suggests that this effect would raise serious questions about the true motives of the refusing physicians and the sufficiency of their reasons. This is particularly the case, she argues, when their actions differentially affect *protected* groups and *protected* rights, namely women's reproductive freedoms. This may be a legitimate concern in the context of physician refusals as they are permitted under the current system, since in that case a discriminatory effect might in fact arise, even if no genuinely discriminatory intent was present. However, given the responsibilities imposed on licensing boards and the fact that other willing physicians will be available under the institutional solution, any potential for disparate impact will be lessened (if not eliminated) and any truly discriminatory grounds for refusal can be denied protection, as we will see now.

Impermissible Objections to the Patient
Savulescu worries that the "door to 'value-driven' medicine is a door to a Pandora's box of idiosyncratic, bigoted, discriminatory medicine".[21] However, opening one door does not automatically require opening the other, for not all values must be accepted as valid in the professional setting, even if people should generally be permitted to live according to their own moral convictions. Instead, a coherent and bright line can be drawn to prevent things from going too far—specifically, physician refusals grounded in invidious discrimination without concern for

particularized circumstances should be rejected by licensing boards.[22] In fact, many laws of general application already prohibit such discrimination at both the state and federal level.[23]

Nevertheless, there is still significant room for normative concern. Should physicians with moral objections that conflict with these existing anti-discrimination laws be exempted from their proscriptions (even if that is not currently the case)? And should physicians engaging in a particular form of discrimination on moral grounds that is not prohibited by any applicable law be permitted to freely continue such behavior? The answer to both questions seems clearly to be no, for in these cases, there is serious dignitary harm beyond the mere offensiveness of having someone disagree with or judge your choice to seek a particular service. Instead, in this context, the refuser is condemning *who* you are.

The entitlement to employ one's own discretion in deciding with whom to associate is widely considered an important attribute of professionalism.[24] From both a legal and professional perspective, however, this entitlement does not generally encompass discretion to discriminate on the basis of wholly irrelevant characteristics. For example, the American Medical Association's Code of Medical Ethics permits doctors to refuse services that are contrary to their moral or religious beliefs, but prohibits invidious discrimination.[25] Similarly, the Declaration of Geneva prohibits physicians from allowing "considerations of age, disease or disability, creed, ethnic origin, gender, nationality, political affiliation, race, sexual orientation, social standing or any other factor" to influence the fulfillment of their duties to patients.[26] Further, many existing refusal statutes contemplate consistent refusals to perform a procedure for any patient, not only for certain patients in a discriminatory fashion.[27] In fact, Mississippi's conscience clause explicitly withholds protection from physicians engaged in discriminatory practices, eliminating any confusion as to whether licensing boards retain the freedom, broad refusal protection notwithstanding, to discipline such doctors.[28]

Although this distinction may appear to rely on an arbitrary boundary, since physicians may have moral objections to the patient that are equally strong and real as the objections others have to the service itself, the validity of conscientious objections is, by necessity, socially defined. For example, a man subject to the military draft may be as genuinely opposed to a particular war as another man is to the idea of war in general, but only the latter is treated as having a valid reason for refusing to serve.[29] Similarly, a racist or anti-Semite might claim that donning his Ku Klux Klan robes or swastikas is as important as a Jewish man believes wearing his yarmulke is. However, most people see no problem allowing the Jewish man to wear his yarmulke to work, while many of those same people would raise serious objections if the racist or anti-Semite appeared at the office wearing his offensive paraphernalia. As these examples demonstrate, we have decided to respect some types of conscience in the public sphere and not others, regardless of genuineness, often based on their social consequences—and the social consequences of discrimination in medicine are very real.[30]

Allowing physicians to discriminate based on race, gender, age, sexual orientation, or any other physical characteristic, when these things have nothing at all to do with the factors properly taken into account by a physician, will sully the reputation of the medical profession and diminish social trust in doctors. Physicians are expected, as professionals, to be model citizens and to behave exceptionally well. Why, then, would we allow them to discriminate on invidious grounds when we do not allow barbers, restaurateurs, or landlords to do so? It may be true that "the nature of the doctor-patient relationship is fundamentally more intimate than the sorts of interactions that occur between landlords and tenants or innkeepers and guests,"[31] but because invidious discrimination degrades health care no matter what the rationale, a doctor's clinical decisions must remain entirely uninfluenced by judgments about a patient's worth.

As Thomas May has noted, we no longer accept separate but equal in this country—and of course, we never should have. This explains why we must continue to be concerned about the grounds for a physician's refusal of care even if similar care remains available from a willing provider accessible to the patient.[32] Because "both good and evil can emanate from conscience,"[33] and invidious discrimination is clearly an evil that can result from otherwise genuine and potentially even nonmalicious conscientious beliefs, it should not be protected by licensing boards. Importantly, even if a patient has no right per se to a given medical service, if the physician is willing to provide that service to others who are the same in all relevant respects, it seems that the patient has at least a de facto moral right to nondiscrimination—and in most situations, a legal right against discrimination as well. Patients with similar clinical characteristics absolutely ought to be treated similarly.

Further, it is essential to note that as a matter of social negotiation, physicians are expected to treat even the "bad guys"—including the abusive husband, the child molester, the terrorist, the murderer, the woman injured by the very bomb she detonated at an abortion clinic, and even medical malpractice lawyers.[34] These are somewhat more difficult cases than refusal to treat the black patient, for example, since refusals of these patients would in fact rest on denunciation of the patient's choices. Nevertheless, such refusals should not be accepted as valid because they are grounded solely in the judgment that the patient is somehow immoral or unworthy of respect, abstracted from the specific care the patient is now seeking; in these cases, too, the social harm of allowing conscientious refusal truly outweighs the benefits. It is the repugnance, dislike, loathing, or contempt expressed by the refuser for the patient him or herself based on his or her physical characteristics or broad life choices that is singularly impermissible as a ground for conscientious refusal.[35] Critically, this motive is not reflected in a determination that a service is inappropriate for a particular patient, as in the breast implant or futile care examples offered above, nor in an objection to a particular medical service.

Thus, the widely publicized refusal by a California physician to treat a little girl's ear infection because her mother failed to conceal her own tattoos is clearly inappropriate on several levels, and is not redeemed by the fact that the doctor notified the mother of his discriminatory policy when she made the appointment.[36] Likewise, a physician may truly believe the homosexual lifestyle to be sinful, but if he is willing to implant donor embryos in a heterosexual woman who is unable to conceive naturally due to her age and not due to any deviation from "species-typical functioning" in either herself or her partner, he cannot be permitted to register a refusal to implant embryos in a homosexual woman unable to conceive naturally both due to her age and her sexual orientation.[37] In this example, the patient's sexual orientation is completely irrelevant to the reason for infertility and to the projected medical outcome of the treatment—and, as an increasing amount of research demonstrates, to the social adjustment of the resultant child.[38]. The bottom line is that this doctor clearly has no objection to the service standing alone, but rather would be expressing an invalid moral objection to the homosexual patient.[39]

Nevertheless, it appears that many physicians may be engaging in precisely this sort of impermissible discrimination. As of 2005, 53 percent of fertility clinics responding to a national survey stated that they would be very or extremely likely to turn away a single man, and 20 percent said they would turn away a single woman. Forty-eight percent reported that they would turn away a gay male couple, while 17 percent said they would turn away a lesbian couple.[40] Not only is this data troubling given the gender disparities represented, but it also stands squarely against the position taken by the American Society for Reproductive Medicine on this issue. While recognizing that even though some people genuinely view homosexuality as immoral and oppose facilitating gay and lesbian reproduction, ASRM maintains that

the ethical duty to treat persons with equal respect requires that fertility programs treat single persons and gay and lesbian couples equally with married couples in determining which services to provide. Unless other aspects of the situation also would disqualify married or heterosexual individuals from services, such as serious doubts about whether they will be fit or responsible child rearers or the fact that the program does not offer anyone a desired service, for example, gestational surrogacy, we find no sound ethical basis for licensed professionals to deny reproductive services to unmarried or homosexual persons.[41]

Importantly, at least one court has refused to allow the religious beliefs of physicians to be used to deny services on a discriminatory basis, although at the time this book went to press, its decision was still under review by the California Supreme Court.[42] The somewhat convoluted case involves the denial of artificial insemination services to Lupita Benitez, an unmarried lesbian woman, by doctors at a Southern California infertility clinic.[43] Benitez alleges that her doctor disclosed religious objections to helping her conceive by means of artificial insemination, but claims that the physician nevertheless agreed to provide fertility services and told her that it would not be a problem for another physician in the office to perform the actual insemination procedure.[44] Thus, Benitez underwent months of preparatory fertility treatment without any indication that she should seek an alternative provider before the clinic's doctors finally refused to proceed.[45] Ultimately, Benitez was forced to obtain assistance from a physician outside of her health plan, incurring additional time burdens and expenses before giving birth to a son. The case has clear implications for physician freedom of conscience, although the state's conscience clause protects only refusals to participate in abortion and is not implicated in the litigation,[46] which has been framed more broadly as involving issues of constitutional religious freedoms.

Discrimination based on sexual orientation is barred by California's Unruh Civil Rights Act. Marital-status discrimination has also been forbidden by case law, which has now been codified in the statute, though these developments occurred after the doctors in this case refused to provide the requested service to Benitez.[47] Because judicial decisions are generally given retroactive effect, the trial court held that the doctors had violated the statute regardless of whether their discriminatory behavior was based on Benitez's marital status or sexual orientation, and refused to allow the doctors' religious beliefs to be used as a defense.

However, the intermediate appellate court chose to apply the marital-status change to the Civil Rights Act only prospectively, since it found that the parties had relied on the earlier interpretation of the law.[48] Thus, it held that the trial court had to revisit the issue of whether the doctors had really violated the statute. This in turn would depend on a factual determination of whether their refusal was based on sexual orientation as Benitez claims, which would have been prohibited at the time, or marital status as the physicians claim, which would have been acceptable. Only if it were based on sexual orientation would the trial court need to reach the issue of the physicians' religious beliefs. In that event, the intermediate court held that that the physicians could assert their religious reasons for denying the treatment in front of a jury to determine if they were sincerely held, even though the trial court had previously held that the religious nature of these reasons was immaterial to liability for violation of the statute.[49]

Before the matter returned to the lower court, however, the California Supreme Court agreed to hear Benitez's appeal. This court will focus only on the physicians' defense, and has characterized the matter before it as a question of whether a physician has a "constitutional right to refuse on religious grounds to perform a medical procedure for a patient because of the patient's sexual orientation, or [whether] the provisions of the Unruh Act preclude such discrimination in the provision of services notwithstanding the physician's religious beliefs." The case will unfortunately shed no light on how physician refusals should be handled when they are not invidiously discriminatory or when they are based on general moral, but not religious, beliefs, but despite these limitations, this first case of its kind will be an important contribution to the debate, as far as it goes. Constitutional issues will play the primary role in the ultimate outcome, but as a normative ethical matter, regardless of whether the doctors refused to provide Benitez with artificial insemination due to her marital status or her sexual orientation, and regardless of whether their objections were rooted in their religious beliefs, they behaved improperly. They rejected the patient and not the service, and thus should not be shielded from liability.

Before moving on, it is important to recognize that withholding protection from discriminatory moral refusals may create significant disincentive effects. This is because prohibiting discrimination by physicians would in fact prevent a physician from performing a particular service for any patient if he is unwilling to provide it to all patients, unless he could point to some relevant difference between them.[50] Clearly, this could limit the absolute number of physicians performing the service in question, which is somewhat troublesome given the access concerns raised in the face of permitting conscientious refusal at all. Nevertheless, this appears to be a necessary trade-off to prevent improper discrimination, the presence of which in the profession does nothing to advance legitimate moral debate and the avoidance of which must be a clear professional obligation of each individual physician.

Tests of Sincerity

While the lines of validity have been drawn with regard to the sorts of conscientious refusals that ought to be accepted, concerns of abuse may linger. What is to prevent the physician whose refusal is based on impermissible grounds from dressing it up as one licensing boards will accept? Clearly, appropriate safeguards will be necessary, but what might they entail? This is an important question, but licensing boards need not start from scratch in developing measures to ensure that refusal registrants do not take advantage of the system. Instead, they can draw on existing judicial experience testing the sincerity of claimed conscientious objection to forced military service.[51]

The Supreme Court's decisions in *U.S. v. Seeger*[52] and *Welsh v. U.S.*[53] provide a combination of factors for testing the authenticity of moral reasons for conscientious objection. The easiest test is adherence to an established religion the tenets of which prohibit the behavior in question, demonstrated by more than merely claiming membership, but rather by actual participation. Barring that, however, the Court will ask whether the alleged ethical standard creating the objection runs so deep into the core of the objector's identity that it would be virtually impossible for him to violate it because the personal consequences would be too great, though this measure of sincerity begins to look more like a test of validity. Finally, the Court may examine whether the objector would be willing to pay some price for the nonconformity in order to show earnestness and depth of belief.[54]

Admittedly, each of these factors is fairly vague and open to interpretation, though they have been given some concrete substance in the actual practice of draft boards.[55] All men must register with the Selective Service on turning eighteen, at which time no conscientious objections are accepted. However, in the event of an actual military draft, those registrants who wish to object can file a claim with the Selective Service and "must establish to the satisfaction of the board that [they] are conscientiously opposed to participation in [combatant or noncombatant] military training and service in any war, based on deeply held moral, ethical or religious beliefs." To do this, objectors must respond to three queries, which could be easily made applicable to physician refusers, as demonstrated by the minimal additions in brackets:

1. Describe your beliefs, which are the reasons for you claiming conscientious objection to combatant military training and service or to all military training and service [or, which are the reasons for you registering conscientious refusal to provide X medical service].

2. Describe how and when you acquired these beliefs. Your answer may include such information as the influence of family members or other persons; training, if applicable; your personal experiences; membership in organizations; books and readings, which influenced you.

3. Explain what most clearly shows that your beliefs are deeply held. You may wish to include a description of how your beliefs affect the way you live.[56]

Objectors are encouraged to attach letters of support from those familiar with their beliefs, such as family members, friends, teachers, and clergy, as well as any other supportive documentation, perhaps indicating examples of other behavior consistent with the objectors' claimed beliefs.

Importantly, objectors must certify the truth and accuracy of everything they submit, and are threatened with both imprisonment and financial penalties if they are found to have attempted to evade the draft laws. In the medical context, boards could threaten physicians with loss of their license for registering refusals on insincere grounds. If the draft board is convinced of the sincerity of the registrant's acceptable objections, he is exempt from the draft. If the board remains unconvinced, however, the registrant can appeal, though his refusal may eventually be rejected, in which case, he must face the draft or prison—or in the physician's case, provision of the service or lack of protection from the consequences.

Learning from the experience of the Selective Service system, licensing boards can tailor their tests, review processes, and consequences as they see fit, likely based on cost-benefit analyses of weeding out those doctors unprofessionally trying to game the system. It is important to note, however, that boards should not adopt their conscientious objection standards wholesale from the military context, since the lines of validity drawn there may be narrower than those established here for the medical setting. In particular, while many physicians' grounds for refusal are in fact so deeply held and integral to their identity that they would choose to leave the profession rather than violate them, a physician's moral objection may be genuine and worthy of protection even if it is not as strong.[57] This weaker objection, however, might be rejected under Supreme Court jurisprudence regarding the draft. Additionally, while some commentators have suggested that acceptable conscientious refusals by physicians must be consistent with the refuser's other beliefs and actions,[58] boards would have to tread carefully in their consistency requirements in order to allow physicians whose beliefs have genuinely changed to register refusal.

Testing the sincerity of another person's internal beliefs will always risk imperfection, though refusing physicians, like other conscientious objectors, should bear the burden of showing adherence to the convictions used to support their refusal.[59] A sufficiently devious, unprofessional, and morally reprehensible physician could probably succeed in evading the boards' regulations, getting away with protection for refusals that are actually based on impermissible grounds unworthy of protection by falsely characterizing them in such a way that licensing boards will not balk. However, given the fact that physicians have an obligation to comply with state regulatory standards and the fact that moral objections exist that are truly worthy of protection, mere potential for abuse or error is certainly no reason to ban conscience clauses altogether. Many laws are incapable of perfect drafting that would narrowly protect only those in genuine need of protection and never extend to others, but this imperfection is often tolerable and preferable to the alternatives, as it is here.

We can also take heart in the fact that while conscience indeed "can be an excuse for vice,"[60] the most highly publicized cases of conscientious refusal have involved health-care professionals with genuine moral objections to the tasks requested of them. Notably, this trend developed in spite of the vagueness and breadth of existing conscience clauses that impose absolutely no tests of refuser sincerity; with the additional safeguards proposed in this section, abuse is even less likely. Further, even if some physicians get away with improper refusals, patients will not be significantly harmed by this breach of professional obligation, since under the institutional solution, they should be able to obtain the desired service from another provider with relative ease. Reasons do matter, but the bigger question is about consequences. In this context, with these standards and safeguards, negative consequences can be drastically minimized, if not avoided entirely.

As this discussion demonstrates, measuring the supply of willing physicians in order to assess whether it is sufficient to meet patient demand for services could be achieved by simply having physicians register their refusals with the licensing board, though they need not register a refusal to provide any service determined to be beyond the scope of the board's obligation. Of course, deciding which refusals should be accepted is far more complex. Certainly, a physician's reasons for refusal must be genuinely held, and licensing boards can draw on existing models for testing the sincerity of conscientious objection to give teeth to this requirement.

Determinations regarding the validity of a physician's reasons present greater difficulty and will require that licensing boards set standards governing precisely which value judgments a physician can appropriately bring into the clinical encounter. Given the importance of moral diversity and freedom of conscience, boards ought to presume that the physician's grounds for refusal are acceptable, unless there is a strong reason to reject them. Thus, boards should accept any ethically grounded refusals based on objections to particular medical services abstracted from the patient requesting them, since in these cases the physician is simply expressing disagreement as to the proper ends of medicine. He would not be making any judgment about the patient per se, but rather would be judging "the morality of the *acts* required by the procedure the patient seeks."[61]

However, refusals based on wholly irrelevant patient characteristics, such as race, must be denied, though not all patient-based refusals should be rejected, particularly when they are the result of individualized reasoning about whether the requested intervention would truly be good for the patient and are not the result of simple invidious discrimination. While the access concerns surrounding refusal might be resolved regardless of the grounds for refusal, perhaps suggesting that we ought to ignore reasons and offer blind protection to all claims of conscience, that approach would placate only one type of opposition to conscientious refusal by physicians. Egalitarian justice concerns are also important, which is why discriminatory reasons simply cannot be endorsed.[62]

These specifications of validity will help mitigate the current ambiguity of existing statutes with regard to their definitions of conscience, raising the refusal bar and demanding more precise justifications for refusal to comply with patient requests. Although most important ethical questions fail to lend themselves to resolution by simple formulas, the distinctions described here avoid blanket acceptance of appeals to conscience and are thus far from simple. Instead, they are based on a nuanced analysis of relevant values, interests, and likely consequences, but are simultaneously clear, avoid unduly strict limits on the exercise of physician conscience, and do not rely on contested ethical standards.[63]

7

Calibrating Supply and Demand

While the tasks of determining which patient access demands licensing boards should be responsible for meeting and which physician refusals they should be willing to accept present some difficult normative questions, other definitional and measurement concerns remain. In order to know where to measure supply and demand and to determine whether the profession is sufficiently addressing those patient access concerns stemming from conscientious refusal, geographic boundaries must be established. If patients demand abortions, for example, how far is too far to expect women to go to find a willing provider before we will say that the licensing board has unacceptably failed to meet its obligations? Clearly, it is not sufficient for boards to simply ensure that there are enough providers within the entire state to meet patient demand, since state-level availability may turn out to mean a concentration of willing physicians in urban areas with no access at all for patients unable to get there. A finer-tuned catchment area is needed, but how narrow or wide a circle to draw is a matter of reasonable debate. Importantly, however, development of such boundaries is not foreign to the health-care context and licensing boards will be able to look to several analogous situations for guidance.

Once these lines are drawn, we will have a concrete measure of the profession's responsibilities with regard to conflicts of conscience and it will be possible to determine whether those duties have been breached—that is, when a patient is forced to go beyond the decided-on geographic boundary to obtain care as a result of conscientious refusals by doctors within that boundary. In the event such a problem is discovered, boards must remedy the mismatch between physician supply and patient demand that it represents, and will have at their disposal several incentive-based measures by which to do so.

How Far Is Too Far When It Comes to Reasonable Access?

As we saw previously, many commentators would allow room for individual physician conscience so long as there is an alternative source of care, and in the pharmacy context, proposals that require a patient's prescription to be filled by someone in the original pharmacy during the original visit have been relatively well received. Unfortunately, however, commentators generally fail to provide operationalizable details regarding whether any greater burdens would count as unreasonable for the patient, and thus fail to give significant meaning to the requirement that an alternative provider be available beyond the clear case. For example, Elizabeth Fenton and Loren Lomasky suggest that where patients can "easily avail" themselves of other means of satisfying their medical demands, conscientious refusal is acceptable, as it is when providers "abound," though they argue that it is impermissible where such refusal would create "significant hardship" for the patient.[1] But what do all of these broad and vague standards actually mean? The details must be addressed.

Before determining how licensing boards might go about setting the relevant geographic boundaries, however, it is important to note that reasonable access and unreasonable hardship need not mean the same thing for every medical service for which the board bears responsibility. Boards can properly place a greater burden on the patient for some services than for others, allowing the geographic area in which it must be accessible regardless of conscientious refusal to be drawn more broadly or narrowly depending on the circumstances.

To make these distinctions, licensing boards might utilize some of the dividing lines that were deemed inappropriate for determining which demands they have a responsibility to meet. For example, while the medical profession has an obligation to ensure access to services within its control regardless of whether those services meet patient needs or mere patient preferences, patients can legitimately be expected to travel farther to obtain elective procedures than to access those procedures more central to their health. Patients can also be expected to travel farther for nonemergency services, although the repetition of services needed for chronic conditions might counsel toward drawing the applicable geographic circle for those services narrowly, as would the fact that waiting for many types of treatment might involve significant worry, inconvenience, discomfort, and potentially increased risk.[2]

Licensing boards should also consider the emotional factors facing patients seeking certain types of medical services. For example, given the emotional trauma of abortion for many patients and the likely emotional state of the patient when making this choice, it may be best to minimize the burden on women seeking this procedure, even when it is neither a medical necessity nor particularly time-sensitive. Outside of the abortion context, at least one court has given weight to the psychological burden of forcing a patient to seek care elsewhere. Recall the case of *In re Requena*, where the court paid heed to the wishes of a patient suffering from Lou Gehrig's disease to remain in the hospital where she had resided for several months while she refused life-sustaining artificial nutrition and hydration, against the wishes of the hospital and its staff. Despite the fact that a willing hospital was located only seventeen miles away, the court allowed the patient to stay put because it noted that it would be "emotionally and psychologically upsetting to be forced to leave the Hospital."[3]

The court's opinion also noted that asking the patient to go elsewhere would "have significant elements of rejection and casting out which would be burdensome for [her]."[4] However, as we saw in the last chapter, this is exactly the sort of objection to a service that ought to be permitted. Perhaps seventeen miles was too far in this context, especially if it would have prevented the patient's family from visiting as frequently during such a crucial time, but if an alternative willing provider had been reasonably accessible, the dignitary harms associated with feelings of rejection are insufficient grounds to force a violation of conscience. The court implicitly acknowledged this, given that it was willing to allow some health-care professionals to refuse so long as someone in the hospital would provide the needed care. Under those circumstances, feelings of rejection could certainly remain, so access concerns were clearly the court's primary focus. While it would go too far to require that all patients be able to access care at the precise time and place of their choosing, their needs *and* their preferences both remain important considerations.

Additionally, licensing boards will need to pay close attention to a variety of barriers to access, especially transportation and financial obstacles. As Jill C. Morrison, senior counsel to the National Women's Law Center, has aptly noted, "people couldn't drive out of Louisiana to get out of ... hurricane [Katrina]; people can't drive 40, 50 miles to the next nearest pharmacy, especially if they don't have a car."[5] Even if they do, not everyone is able to take the time off from work or secure the child care that potentially may be necessary in order to seek medical services that are only available a significant distance away. Ensuring meaningful, not merely nominal, availability is paramount. Therefore, boards must focus on actual travel patterns rather than on artificial boundaries like state, county, city, or zip-code borders. Not all physicians within those contrived catchment areas will be available to all residents, and physicians at the edges will often be accessible to outsiders.[6]

These factors suggest that licensing boards have a great deal to consider when drawing relevant geographic borders in which to define their responsibilities to mitigate the impact of conscientious refusals, and by the end of this chapter, we will see that a number of other crucial issues come into play as well. Therefore, just as these boards must gather diverse perspectives when determining which patient demands they have a responsibility to meet, they must also consult laypersons and experts outside of the medical profession, such as economists and patient advocates, when defining these important geographic constraints, which should likely be vetted through notice-and-comment rulemaking as well. While it is clearly a complicated task to operationalize how much burden is too great to impose on patients seeking a variety of medical services from willing physicians, licensing bodies will not have to reinvent the wheel when assessing appropriate boundaries within which demand for services should be measured and satisfied. Instead, boards will be able to draw from principles used to define relevant markets in antitrust law, certificate-of-need requirements for the establishment of new or expanded health-care facilities, and federal programs designed to improve access to care in underserved areas.

In antitrust law, courts must define the relevant markets in order to determine whether there is an illegal monopoly at work, since monopoly power is characterized as the market share of a given firm in a given market being sufficient to control prices and/or exclude competition.[7] The definition of markets used in this context could also be useful in determining the relevant geographic area in which patients can reasonably be expected to seek a willing physician.

The antitrust market consists of two distinct, but related, elements: the product market and the geographic market.[8] Assessing these factors together establishes the smallest area in which the elasticity of demand and supply is sufficiently low that a supplier could profitably reduce output and introduce a "small but significant and nontransitory" increase in price without customers seeking the product elsewhere or seeking an alternative product altogether.[9] Translating to the health-care setting, then, licensing boards could ask how many willing physicians are necessary in a given area before patients will forgo the service, seek a substitute, or be forced to find a provider outside of the bounds of the geographic market. That is the number boards will have the obligation to maintain access to, unless, as we will see, these effects on patient behavior occur not as a result of conscientious objection on its own, but rather as a result of additional factors contributing to access concerns.

To determine the relevant product or service market, one must ask whether the products or services in question are reasonably interchangeable by consumers for the same purposes. In antitrust cases, courts determine interchangeability by assessing customer views on the matter, price, cross-elasticity of demand, perception of the relevant industry regarding the scope of its market, the firm's own view of who its competitors are, and differentiation in the consumer market.[10] With regard to the services prone to raise issues of conscience for providers, the possibility of patients substituting one service for another is relatively unlikely and determining the service market is therefore straightforward. For example, while there are different methods of surgical abortion, when a woman has reached the point of gestation at which mifepristone (the "abortion pill," RU486) is no longer an option—about eight weeks— she really has no alternative to some variation of surgical abortion should she desire to terminate the pregnancy. The service market is clear.

Licensing boards are likely to find the antitrust discussion of geographic markets more relevant for their purposes, since these are based on the area in which the seller currently operates and effectively competes with others, as well as the region to which consumers could practicably go, though may not actually go at this time, to obtain the service should circumstances change.[11] Similarly, where patients can realistically access the services physicians might refuse to provide on grounds of conscience is at the heart of the conscience clause debate. However, there is substantial disagreement among economists as to how the relevant geographic market should be determined, and unfortunately, there is no standardized, bright-line measure. Indeed, experts reviewing the same set of data frequently reach different conclusions. Part of the difficulty is created by the fact that the question of geographic markets is not static. It is concerned not only with what patients currently do, but also asks what patients might be willing to do, and predictions are always more difficult and less reliable than hard measurements of current facts.[12] Nevertheless, multiple tools are available to delineate antitrust geographic markets and may be particularly useful when employed in tandem.

Patient-origin data are helpful at answering at least part of the question: How do patients behave now? This data is often collected in the hospital setting by state agencies and hospital associations, so it is readily attainable, though maybe not for individual physician practices. It offers important information about patient willingness to travel for care and their present reactions to various price and quality conditions, and serves as a basis for educated guesses about patient behavior should one or more currently available sources of care be eliminated.[13] Courts will also generally consider the raw distance traveled by patients for healthcare services, actual travel times, transportation costs, physician admittance patterns, and the views of market participants themselves.[14]

Additionally, antitrust experts have offered several other tests to supplement patient-origin data. For example, the direct-competitor test measures the overlap between the primary service areas of available providers, generally defined as the smallest set of zip codes from which the providers in question draw 90 percent of their patients. A large overlap indicates that patients can, and are willing to, travel between service areas, suggesting that the relevant geographic market includes all of the overlapping areas. On the other hand, little overlap demonstrates only that patients do not presently travel between the service areas for care, not that they are unable or unwilling to do so.[15]

Another option is the critical-loss test, which is based on Department of Justice/Federal Trade Commission Merger Guidelines and establishes the geographic market by assessing whether a hypothetical cost increase of 5 percent would be profitable or whether it would result in the loss of a critical number of patients who would seek care elsewhere. If it would be profitable—that is, patients would not seek alternative providers—then those alternative providers are treated as beyond the geographic market.[16] Finally, the Elzinga-Hogarty test is a well-known standby in geographic-market delineation that examines the percentage of patients in the area under consideration who deal with providers outside of that area and the percentage of patients in the outside area dealing with providers inside.[17] The fewer patients who are traveling outside the given market to obtain medical services, the better the indication that services beyond that market boundary are not practically available, suggesting a strong geographic market, but not conclusively demonstrating that outside services are impossible to obtain should the patients need to venture that far. To clarify, in the conscientious-refusal context, the boundaries established by each of these measures would indicate that licensing boards should measure demand for the service within that boundary and ensure access to a sufficient supply of willing physicians therein.

Obviously, these measures would require modifications for the licensing board's purposes, particularly since the tests were developed for application to fungible markets where all consumer migration is solely attributable to price.[18] Our concern is not price, however, but is rather moral choice and geographic access to willing providers. Further, the focus on migration due to price alone is worrisome because, just as consumers in any market may be driven by nonmonetary factors, that is especially true of patients. Quality is an incredibly important factor in the health-care setting and patients may be willing to travel farther to reach physicians with a better reputation.[19] This is certainly the case for non-routine, potentially life-altering, medical interventions where relatively small variations in physician competence could make a large difference.

As alluded to above, a bigger problem with these antitrust measures of geographic market boundaries is that they provide accurate assessments only of existing, rather than potential, patient behavior, though both are vitally important to the question of how far is too far for acceptable patient access. This focus on current behavior could result in several types of errors. First, it might draw the relevant geographic market too narrowly because it captures only the area where patients presently seek care even if they would be willing and able to travel greater (but still reasonable) distances if services were not so readily available closer to home. This type of error would place a greater burden on licensing boards than necessary, since they would have to ensure access to more willing physicians in a given area than patients actually need, although this would also minimize the burdens on patients resulting from conscientious refusal and may therefore appear acceptable. However, this outcome is troubling in light of the potential for genuine shortages of physicians that will be discussed later in this chapter; in that context, demanding more physicians than are actually necessary may result in some areas being left without access to a more appropriate baseline.

Drawing the geographic market too broadly is also a concern, for it would capture and maintain existing, but avoidable, access problems. If patients are currently prepared to travel farther than should reasonably be expected of them because they are already faced with a depressed number of physicians willing to provide a given service in the relevant geographic range, that readiness to travel should not result in a self-perpetuating baseline that effectively punishes patients for traveling long distances for services that should probably be available closer to home. However, this is exactly what could happen if the distance patients currently travel is used to demarcate geographic markets. In other words, what is practicable in the antitrust world because patients actually do it may not be what is normatively desirable.

This problem would be particularly apparent with regard to abortion services, and as we have seen over and over again, resolving the inability of many patients to obtain practical access to this service will be integral to convincing opponents of conscience clauses to accept this institutional solution. For example, as of 2003, 87 percent of U.S. counties had no identifiable abortion provider, and in nonmetropolitan areas, that statistic rose to 97 percent.[20] A similarly stark example is provided by the fact that as of 2007, the entire state of South Carolina had only three abortion clinics.[21] In 2000, 25 percent of women obtaining abortions traveled at least fifty miles for access, and 8 percent traveled more than one hundred miles.[22]

Clearly, these statistics demonstrate that many women already have to travel a significant distance to obtain abortion services, at increased personal expense, both financial and emotional. This may be in part due to existing conscience clauses allowing physicians to refuse to provide these services and also to the general decline in physicians being trained to provide abortions, not to mention continued threats of violence and harassment facing those physicians who will perform the procedure.[23] Should these figures indicate that since travel is the norm to obtain this service, the patient can be expected to bear a significant burden, or should licensing boards try to remedy the existing failure of the profession to ensure decent availability of this service by not using the current figures as a baseline?

This self-perpetuation quandary is analogous to the "Cellophane Fallacy" in the relevant antitrust literature and is the result of failing to account for existing monopoly power. In *U.S. v. E.I. DuPont de Nemours*,[24] the Supreme Court defined the relevant product market for cellophane not as cellophane products, but rather as flexible packaging materials, since it believed that there was cross-elasticity of demand between cellophane and these other goods. However, this definition of the product market was probably incorrect, given that DuPont was charging the highest possible price for cellophane that would not cause it to lose a substantial number of customers. Therefore, the cross-elasticity of demand might have appeared high not because DuPont had no monopoly power, but rather because it was already charging a monopolist price, thus forcing some customers to be willing to use an inferior, but cheaper, product.[25]

In the medical service context, the fact that there may already be a very low number of providers of a given service in a particular area does not necessarily mean that there is no demand for that service, but rather that there are no consequences for refusing to provide that service or that there are insufficient market incentives to do so. In fact, evidence shows that "at least in noncapitated healthcare markets, an increased supply of medical resources leads to increased utilization," indicating that current utilization levels do not necessarily provide the appropriate baseline.[26] What might be considered the appropriate baseline number of willing physicians will be explored in greater detail subsequently. However, for the time being, note that the level of inconvenience that patients can be properly expected to bear will vary with their geographic location and with the level of inconvenience patients in that area would have to accept to get care even if conscientious refusals were not an issue.

One final type of error could result from the fact that antitrust models are able to confidently measure only current patient behavior. Licensing boards might get things intuitively backward, drawing very wide boundaries for important health-care services, since patients are willing to go to greater lengths to obtain access, and drawing boundaries very narrowly for health-care services most would agree are purely elective, since patients might not be willing to go to such extremes for these things.[27] This behavior is based on patient need and other factors, but given their focus on price-based choices alone, the tests are incapable of the nuance necessary to avoid placing a greater burden than is appropriate on patients experiencing serious medical hardships. Nevertheless, courts and other expert commentators have noted that "medical markets tend to be small and highly concentrated because service delivery is inherently local,"[28] so perhaps concern that this type of error would have a devastating impact is unnecessary.

Market definition is certainly complex, and will require significant contribution by experts, but the existing work done in this area will provide at least a jumping-off point for licensing boards seeking to determine how far patients can reasonably be expected to go to obtain services falling within the board's responsibility. Licensing boards might also glean some useful lessons from an examination of certificate-of-need regulations, which set standards for assessing whether there is sufficient demand for new or expanded health-care facilities to justify going forward with plans to build them. While these regulations are intended to ensure that there is not overcrowding of the health-care market and to keep costs down by reducing unnecessary investment in new facilities and services,[29] the general criteria they implement are also used to reveal which services an area does need. They might similarly be used to ensure that patient demands are met with an adequate supply of willing physicians regardless of conscientious refusal.

For example, the Ohio Administrative Code requires the Director of Health to consider current primary and secondary service areas, travel times and accessibility of a proposed project site to the proposed service-area populations, zip-code-based patient-origin data, and special needs or circumstances of the population proposed to be served, including any unusual demographic characteristics.[30] Notably, the director is also to consider "the effectiveness of the project in meeting the health-related needs of medically underserved groups such as low-income individuals, handicapped individuals and minorities,"[31] as well as "the special needs and circumstances resulting from moral and ethical values and the free exercise of religious rights of health-care facilities administered by religious organizations."[32] The Pennsylvania regulation is primed for patient matching, as it requires consideration of the religious needs of the community to be served, as well as broad consumer preferences.[33]

Florida's certificate-of-need statute demands consideration of the availability, quality, accessibility, and extent of utilization of existing health-care facilities.[34] These factors are elaborated on by administrative regulations that require consideration of the need for various services, particularly the needs of medically underserved groups, and the extent of access to those services, particularly by "low income persons, racial and ethnic minorities, women, handicapped persons, other underserved groups and the elderly."[35] Similarly, New York looks at a particular service area, usually the county in which the new construction is proposed, unless evidence indicates that a different border would be more appropriate, as well as migration patterns for health care in that area and distribution of services in relation to the population's distribution.[36] These examples demonstrate that government agencies currently complete the task of drawing relevant boundaries for patient access, and there is no reason to believe that licensing boards could not do the same.

They also demonstrate that open-ended requirements to ensure patient access in health-care settings are not unique. Certificate-of-need statutes and regulations provide broad standards that leave a significant amount of discretion with regard to how things like accessibility and need should be defined and what types of travel times are acceptable. Similarly, many states have pharmacy regulations demanding that owners or pharmacists in charge maintain stocks sufficient "to meet the normal demands of the trading area or patient base the pharmacy serves,"[37] as well as requiring them to ensure that "a sufficient number of pharmacists and supportive personnel are available to operate such pharmacy competently, safely, and to meet the needs of patients."[38] Terms like trading area, patient base, and sufficient are not further defined, but are rather left to the discretion of the government compliance officer. If legislatures and licensing boards are looking for more specific guidance, however, they may appeal to one additional analogy, perhaps the most relevant to the task at hand.

In 1978, the federal government created a system to identify areas in need of physicians and other health-care providers from the Department of Health and Human Service's National Health Service Corps programs, which were designated health professional shortage areas (HPSAs). This designation is now used by several other federal programs to ensure that there is sufficient access to primary care, dental, and mental health services. It is based on a set of rather specific criteria that importantly take no account of whether a particular geographic area is urban or rural or whether it corresponds to the boundaries of any political subdivision.[39]

The Secretary of HHS will label an area an HPSA for primary care services if each of the following conditions is present:

1. The area is a rational area for the delivery of primary medical care services.
2. One of the following conditions prevails within the area:
(a) The area has a population to full-time-equivalent primary care physician ratio of at least 3,500:1.
(b) The area has a population to full-time-equivalent primary care physician ratio of less than 3,500:1 but greater than 3,000:1 and has unusually high needs for primary care services or insufficient capacity of existing primary care providers.
3. Primary medical care professionals in contiguous areas are overutilized, excessively distant, or inaccessible to the population of the area under consideration.[40]

These criteria have a certain level of arbitrariness built in, as all dividing lines must. In addition to the 3,500:1 ratio, artificial categories are established to define rational areas for the delivery of care, which include any of the following:

(i) A county, or a group of contiguous counties whose population centers are within 30 minutes travel time of each other.

(ii) A portion of a county, or an area made up of portions of more than one county, whose population, because of topography, market or transportation patterns, distinctive population characteristics or other factors, has limited access to contiguous area resources, as measured generally by a travel time greater than 30 minutes to such resources.

(iii) Established neighborhoods and communities within metropolitan areas which display a strong self-identity (as indicated by a homogeneous socioeconomic or demographic structure and/or a tradition of interaction or interdependency), have limited interaction with contiguous areas, and which, in general, have a minimum population of 20,000.[41]

The HPSA regulations appropriately measure access based on travel time rather than strict mileage estimates,[42] and also consider surrogate measures of access, such as wait times for appointments, average waiting times at appointments before seeing the doctor, or abnormally low utilization of health services.[43] They provide standards for determining areas with unusually high needs for services, such as those in which more than 20 percent of the population have incomes below the poverty line.[44] They also consider whether providers in contiguous geographic areas will be available to help alleviate shortages in supply based on whether there are significant differences between the demographic or socioeconomic characteristics of the area under consideration and the contiguous area, such as language barriers, and whether the medical resources of the contiguous area will be financially accessible to residents of the area under consideration.[45]

These categories could be similarly adopted (as is, enlarged, or contracted) to define areas in which licensing boards are responsible for ensuring an adequate number of willing physicians. For example, it may be appropriate to ask patients to travel thirty minutes or more to see a reproductive specialist willing to perform IVF, but perhaps patients should have not have to travel so far for the removal of life-sustaining treatment. Discretion would reside with the state legislature or licensing boards since it is not clear precisely which services are more important than others,[46] but either body could adapt these HPSA regulations to the circumstances at hand.

State legislatures interested in implementing the institutional solution could choose to provide a significant amount of guidance to licensing boards, laying out which factors they must consider when determining the relevant geographic areas in which the supply of willing physicians must adequately meet patient demand for medical services. Alternatively, they might opt to take advantage of the flexibility offered by an agency approach, providing licensing boards with a significant amount of discretion in measuring the relevant geographic area, as well as determining precisely when patient demand is sufficiently satisfied. Either way, there are several existing models to offer guidance on these questions.

However, if none of the paradigms described above are helpful, licensing bodies, together with the state legislature, professional organizations, patient-advocacy groups, and other interested stakeholders, might as a last resort negotiate some sort of arbitrary geographic radius based on demand for the service at issue, degree of need for that service, the supply of physicians available for recruitment, and the existence of supportive institutions, such as hospitals and specialty medical centers. The bottom line is that such line-drawing is feasible as demonstrated by the fact that government agencies are already regularly expected to engage in these sorts of measurements in other contexts.

Ensuring and Preserving Reasonable Access

Once state legislatures and/or licensing boards have overcome the measurement and definitional challenges before them, they must undertake their most important task of all—endeavoring to remedy patient access problems stemming from conscientious refusals in the relevant geographic areas. Patient demand for services and the supply of physicians willing to provide them may fortuitously be in line without any licensing board intervention, particularly in urban locales, such that patients can access a desired service from a willing doctor despite the fact that some physicians in that area would conscientiously refuse to provide it. However, when intervention is required because there are not enough willing providers in the relevant area, licensing boards will have a variety of tools at their disposal.

Market Failures

There are two ways in which supply and demand may potentially be misaligned as a result of conscientious refusal that state licensing boards will have an obligation to remedy. The first involves, for lack of a better term, the *artificial* "last doctor in town," a situation that arises when an area is populated with a significant number of physicians, but is morally homogeneous such that all of the area's physicians refuse to perform the same medical services, such as abortion. If there is sufficient diversity within the profession, the moral refusals of physicians will not always overlap in a way that denies availability and creates this sort of problem. Thus, in order to deal with the artificial last-doctor scenario, licensing boards will need to take steps to encourage and facilitate the recruitment and long-term maintenance of physicians with moral views complementary to those of physicians already practicing in the relevant geographic area. Just as some commentators have argued that the legal profession should encourage the "cultivation, financial support, and geographic deployment of lawyers with sincere commitments to certain types of unpopular representation to meet the need for that representation where it arises,"[47] licensing boards must do the same.

Licensing boards might also be faced with *actual* last-doctor-in-town scenarios, resulting not from the overlapping refusals of many physicians, but rather from a refusal by the lone physician in the relevant geographic area or the complete absence of any physician with the technical competence to perform the requested service. This problem cannot be solved solely by encouraging moral diversity, but also demands concerted efforts to draw ready, willing, and able physicians to the area in the first place. Clearly, licensing boards charged with remedying unbalanced supply and demand will need mechanisms to both improve diversity and increase the absolute numbers of physicians practicing in shortage areas.

Some critics have taken issue with this approach, however, charging refusing physicians with free riding on the new recruits, and arguing that asking one's colleagues "to do what you do not wish to do places an unfair portion of the burden for fulfilling the profession's promise to society on select members."[48] For example, Christopher Meyers and Robert D. Woods suggest that, just as conscientious objectors to the draft are compelled to make some meaningful contribution to the maintenance of the national health, safety, and interest through the provision of non-lethal support for the war effort, refusing physicians ought to be required to provide some other service "that is both of comparable social benefit and, to most physicians, similarly 'distasteful' [to the service they refuse to provide] . . . [such as] twenty hours per year of indigent medical services."[49] However, this suggestion ignores the fact that refusing physicians do pull their own weight in helping the profession satisfy its social responsibilities. It is not as though they are refusing to do anything at all—they provide some medical services that patients desire and they contribute to the important moral diversity that all members of the community should seek to preserve and promote. More importantly, it is worth repeating that not every physician must provide every service within his or her specialty in order to meet professional obligations.

Physicians are free to design and limit their own practices for a variety of reasons, such as raising a family or simply preferring some types of procedures and patient interactions to others. In fact, even if physicians selected specialties simply to avoid controversy, protests, and boycotts, rather than because they have a genuine moral objection to the services that engender such reactions, they would be allowed to do so and would not be asked to compensate those who do deal with such tribulations. If these professionals are seen as fulfilling their obligations to society even without the imposition of additional burdens, there is no reason that conscientious refusers should be seen any differently. Asking them to perform "distasteful" services as payment for accepting their refusal is inappropriate. When a physician refuses to provide a service on grounds of moral repugnance, he is not necessarily free riding on willing doctors any more than a factory worker is free riding on the efforts of those working at another point along the assembly line, or than a police officer who chooses to work in a community with a relatively low crime rate is free riding on those who take the job in more dangerous metropolitan areas.[50] Instead, the physician is acceptably passing the patient on to someone who is more capable of providing the service without the impairment of moral conflict or jeopardizing his moral integrity, while the refuser helps patients in other ways. Aside from the requirements discussed in chapter 9, no more should be demanded of him.

The free-rider issue aside, one more important question is why either type of last-doctor-in-town scenario arises at all, given that the market is normally capable of appropriately calibrating supply and demand. In other words, if there is a demand for physicians to provide abortions or assisted suicide (where legal) or sex selection in areas where existing doctors refuse to perform these services, why aren't willing physicians naturally drawn there? The most direct answer is that physicians tend to be affluent and, in many areas, are in short supply as a result of heavy regulation of the number of available medical school and residency slots, giving them the freedom to locate where they prefer to live without having to sacrifice the ability to make an excellent living. There may be a significant number of patients in location X, for example, but if all of the doctors in location Y have a sufficient number of patients to occupy their time, why leave?[51] Further, specialists generally settle in urban areas,[52] and there is reason to believe that moral shortages overlap with rural shortages given that areas with fewer physicians are more likely to run into the artificial or actual last-doctor-in-town problems. If this is true, we can hypothesize that since unaided, natural market forces appear unlikely to resolve the scarcity of physicians in rural areas,[53] the same is true for areas experiencing a scarcity of morally willing physicians.

Regardless of demand for rural physicians, a variety of factors conspire to keep doctors put in more urban, or at least suburban, environments. While students from rural areas are likely to return there on completion of their residency, fewer people from rural areas attend medical school. In addition, most teaching institutions are located in metropolitan areas, resulting in students and medical residents becoming more accustomed to that environment and choosing to settle there. Physicians also tend to marry other professionals and chances of both partners finding satisfactory careers are higher in urban areas where job opportunities are more plentiful and diverse. Finally, physicians may become isolated in rural practice, rendering it more demanding despite the fact that rural doctors earn less.[54]

While conscientious refusal shortages are also possible in morally homogeneous metropolitan areas, perhaps like Salt Lake City, many of the factors contributing to these problems would be amenable to improvement through market incentives offered by licensing boards. Others have noted that targeted incentives offer a powerful mechanism to improve the willingness of health professionals to practice in rural areas. In fact, that is precisely what we have seen in states like Texas, where new statutory limits on malpractice damages have lured a significant number of out-of-state physicians into relocating there, many of whom are filling shortages in previously underserved areas.[55] Of course, this may not be the best way to recruit physicians, for if malpractice caps are set too low, patients may be left vulnerable to poor care. Nevertheless, the development of economic motivators is promising given that physicians, at least those who are morally willing to provide a particular medical service, are likely to act as rational economic beings. So which mechanisms might licensing boards have in their toolbox—not to pay off conscientious refusers, but instead, to draw in willing providers?

Mechanisms to Resolve Access Problems
Just as licensing boards had a variety of sources of guidance for their other novel undertakings, the same is true for the development of economic incentives and other tools to calibrate the supply of physicians and the demand for services only they can provide. Boards could begin by drawing on interventions directed at ensuring sufficient availability of care in rural areas, adapting them to shortage areas created by conscientious refusal. For example, the government agency responsible for designating health professional shortage areas awards scholarships and provides educational loan repayments or forgiveness to students and health professionals in exchange for a commitment to practice in HPSAs for at least two years.[56] In addition to these benefits, physicians receive special bonuses for services provided to Medicare beneficiaries residing in those areas.[57] Licensing boards, perhaps in conjunction with medical schools and professional organizations, could implement similar programs for students and physicians willing to provide needed and unpopular services in analogous "moral diversity shortage areas." This very simple approach is likely to be quite effective.

Other mechanisms addressed at rural shortages may look appealing as well, but we must be wary of their potential to exclude conscientious refusers from the profession by imposing the sort of restrictions at the front end before individuals have become doctors that we have already argued against imposing at the point of licensure or thereafter. In this regard, note that the Council on Graduate Medical Education has recognized that "one of the most powerful ways to remedy problems of rural geographic maldistribution is to change the medical education system so that it *selects*, trains, and deploys more health-care workers who choose to practice in rural areas.... To the extent that we train health professionals who prefer rural practice over other alternatives, it may be possible to improve physician distribution without the need to create special delivery systems or invoke some element of coercion in location choice."[58] Some medical schools have accepted this advice, implementing programs that recruit and selectively admit applicants raised in rural areas who intend to practice family medicine in underserved locations.[59]

Encouraged by the fact that selective recruitment is the most important determinant of students' plans and eventual areas of practice,[60] licensing boards may seek to work with medical schools to develop similar programs aimed at applicants willing to perform controversial services in areas of need. However, even if such willingness was used as nothing more than a "plus" factor in the admissions process, it would still have the predictable effect of making it more difficult for conscientious refusers to gain admission to medical schools or residency programs, for there would simply be fewer spaces available to them.[61] In fact, if this effect did not arise, the program would have failed in its purpose of reducing access problems. While refusers would not be directly rejected for their moral beliefs, the outcome would be the same, thus imposing a serious bar to their entry into the profession; for this reason, what might appear to be a carrot actually turns out to be a stick.

Moreover, assuming that expulsion goes too far, schools would have no way of enforcing such programs unless a student admitted specifically as a result of his or her proclaimed willingness to fill a service gap, but who then fails to do so, is forced to repay prior scholarships or is denied loan reimbursement, if those things were even offered. Rather than limiting admission to medical school, it would be more efficient to simply offer these incentives at the back end to those who actually do help resolve access problems stemming from conscientious refusal. Aside from being more practical, this approach also accounts for the fact that it will be impossible to test the waters to see if patients prefer a certain type of refuser if those refusers are never allowed into medical school in the first place. Further, if medical schools are treated like employers and thus allowed to set the equivalent of bona fide occupational qualifications for their students, they could effectively set the standards for the profession and which services it has an obligation to make available—a task already delegated to licensing boards and interested stakeholders.

Therefore, to retain the benefit of moral diversity by allowing qualified individuals of all stripes to become doctors and take their chances on the market, it is likely best for moral opposition to a particular service not to be used as a factor in excluding individuals from medical school or residency programs. Of course, exceptions could be made for those rare cases where the objectionable service is truly integral to membership in the profession or specialty, perhaps such as the willingness to offer prescription drugs; remember that the objection must at least arguably coincide with the core values of medicine.[62] Meanwhile, boards ought to work to eliminate market failures in other ways through freedom-preserving incentives. When carrots will work, boards should not use sticks. Even when they will not, sticks should be implemented sparingly and with caution because of the social value that conscientious refusers add—encouragement rather than oppression is the preferable model.

Without fear of reintroducing the dangers just averted, licensing boards could nevertheless encourage schools to create targeted efforts to expose students to the access problems some patients are facing and educate students about the profession's responsibilities in hopes of shaping student preferences in a way that corresponds to social needs. Similar programs exist in law schools around the country in the form of legal aid clinics and pro bono programs. Although some students may have always intended to perform precisely this sort of work, others who are exposed to it for the first time in law school may be convinced by their experience to devote themselves to a career of public service. Likewise, medical schools use analogous clinical programs to expose students to inner-city or rural care, and they have had some success in producing physicians willing to practice in those areas.[63] Medical schools could expand these programs to focus on the provision of unpopular services and, whenever possible, the clinics could be housed in areas where conscientious refusal has led to access problems.

Other noncoercive possibilities are also available to help licensing boards mitigate these patient access concerns. For example, boards could work with the federal government to adopt immigration programs aimed at attracting willing international physicians, which is precisely what some federal and state agencies have done in response to the rural health-care worker shortage. The J-1 visa waiver program is used to encourage foreign-born doctors who have just completed their medical education in the United States to stay, trading the two-year foreign residence requirement for at least a three-year service commitment to practice in medically underserved areas.[64] Licensing boards need not limit themselves to American-trained immigrants but could also recruit international medical graduates, especially given evidence that such physicians are equivalent to U.S. medical graduates with regard to skill level and competence.[65] Importantly, however, these recruitment efforts and the resultant brain drain would raise questions of international justice, particularly in light of the shortage of qualified physicians in other countries.[66] Therefore, they should be used by licensing boards only for the most severe shortages and the most serious medical needs, if they are used at all.

A better solution might be available in the form of telemedicine, particularly as the technology improves. Telemedicine refers to the use of electronic information and communications technologies to provide and support health care when distance separates doctors and patients,[67] and it is increasingly recognized as having the ability to mitigate problems associated with the geographic maldistribution of health professionals. Telemedicine transcends spatial and temporal barriers, and in so doing, can provide significant aid to patients and providers in remote and/or underserved areas. While performing surgery with the patient in one place and the doctor in another remains in the realm of science fiction, the Council on Graduate Medical Education commented nearly a decade ago that if a "roving spacecraft can perform atomic spectroscopy on rocks on Mars, there are no conceptual barriers to devising complex interventions at a medical facility 100 miles removed from the base station."[68] In fact, the military is developing this technology to provide care for soldiers on the battlefield, and some academic medical centers and research organizations are also making strides in this regard.[69]

Dealing with an aging population, declining birthrate, and resultant shortage of obstetricians and geographic areas capable of supporting them, Japan has put telemedicine to use as well. Pregnant women in remote areas speak to urban doctors using Internet telephony software, and electronic monitors transmit the baby's heartbeat and mother's vital signs to the doctor, who can assess progress and determine when the time has come to deliver. While the mothers still must travel to their physicians for actual delivery, the use of telemedicine has prevented pregnant women from having to drive long distances only to turn back after false labor, for example.[70]

Telemedicine could conceivably offer at least a partial solution to availability problems resulting from conscientious refusal. To bring it to its full potential, however, licensing boards must loosen restrictions that require a practitioner to obtain a full license, replete with fees, exams, and interview requirements, to deliver telemedical care across state lines.[71] Further, significant advances in robotics must occur before telemedicine will be able to fill gaps in anything other than the provision of consultative services, and it is the physical services themselves that refusers most often find morally objectionable.

Licensing boards might also mitigate access problems by engaging traveling bands of physicians to provide services to areas in which they are unavailable and to which economic incentives and educational changes were not successful in attracting willing physicians. While the American College of Surgeons disapproves of this practice, their reasons are unclear, and evidence suggests that so-called circuit riding and itinerant surgery by doctors based in more populous nearby areas is relatively common and may be increasing.[72] Such a lifestyle may not be the most enticing to physicians who could otherwise settle and make an attractive living by practicing in one place, so licensing boards would certainly have to make it worth their while.

Alternatively, licensing boards could subsidize the travel costs of patients forced to seek care outside the applicable geographic boundaries, which might prove less costly than creating incentives to draw physicians to shortage areas. Unfortunately, however, this solution ignores the fact that financial burdens are not the only improper burdens to impose on patients. As we saw in *In re Requena*, time burdens, as well as the ability to be close to friends and relatives for support through difficult medical procedures, are also legitimate and serious concerns.

Finally, as discussed previously, licensing boards might consider allowing other health-care workers, such as nurse practitioners, midwives, or physician assistants, to perform certain services normally reserved for physicians as a means of improving patient access impaired by conflicts of conscience. If there is a genuine reason for distinguishing between physicians and other health-care professionals on the basis of skill, this approach would be ill-advised, but note that many other providers may be capable of acquiring the skills necessary to provide some of the most controversial services.[73] These providers may assert conscientious refusals of their own, which is why expanding the medical monopoly is not a perfect answer, but it could certainly help since increasing the raw number of competent providers increases the likelihood of patients finding available willing providers. This is particularly important in light of the physician shortages that will be discussed momentarily.

But first we must address the concern that adopting some of these measures to overcome the problems associated with conscientious refusals would amount to unconstitutional favoritism of physicians with particular moral beliefs. However, offering financial incentives, immigration preferences, or special medical school programs is not actual favoritism for those holding certain views, nor is it punishment of those with the opposite beliefs. So long as the government has a legitimate interest in encouraging certain types of behavior and not others, it may offer incentives for the preferred behavior even if the incentive offered affects the ability to exercise some fundamental right. For example, in *Harris v. McRae*, the Supreme Court determined that Congress has a valid interest in protecting potential life, and thus upheld the Hyde Amendment, a law that severely restricts the use of Medicaid funds to pay for abortions and aims to incentivize childbirth by making it the more attractive alternative.[74] A closer analogy for our purposes is provided by *Johnson v. Robison*. In that case, the Supreme Court upheld a program of federal educational benefits intended to incentivize actual combat service that were accordingly offered only to military veterans and not to those conscientious objectors who had provided alternative services.[75]

The objectors argued that the policy interfered with their free exercise rights by increasing the price they had to pay for adherence to their religious beliefs, which is precisely what physicians unwilling to perform the incentivized services may argue in our circumstances. However, the Court explained that the law providing these benefits and limiting them to veterans of combat service was intended to "advance the neutral, secular governmental interests of enhancing military service and aiding the readjustment of military personnel to civilian life. [Conscientious objectors] were not included in this class of beneficiaries, not because of any legislative design to interfere with their free exercise of religion, but because to do so would not rationally promote the Act's purposes."[76] Thus, the Court upheld the law against the free exercise challenge, finding that the denial of the economic value of the veterans' educational benefits was an acceptable incidental burden, if it were truly any burden at all, on the objectors' religious rights.[77]

Not only are the incentives for physician non-refusers one reasonable way to improve patient access—a legitimate, neutral aim that would not be promoted further by offering the benefits to refusers as well—but they would also avoid outright punishment or exclusion of physician refusers as an alternative mechanism of accomplishing that goal. Thus, the proposed interventions are not veiled attempts at government endorsement of particular viewpoints, nor are they meant to disfavor or penalize those who adhere to different beliefs. Rather, they are attempts to protect all sorts of religious and moral values and to shield the very individuals who might claim they are being discriminated against in the receipt of conditional government benefits. Furthermore, boards would still license all those physicians meeting baseline technical and moral competency requirements, incentives aside, so no one is actually harmed or held back.

In effect, physicians with an additional level of "competency"—that is, the willingness to provide needed services—are awarded additional compensation for that bonus. The financial incentives described above might even be viewed as higher payments for members of particular subspecialties, as opposed to compensation for holding certain moral or religious views.[78] A doctor willing to work weekends might be paid more as a result of the added burden of that commitment and the need to satisfy patient demand for weekend services, even if another doctor could not work those shifts because she must attend religious services during that time.[79] By analogy, there should be no problem with compensating physicians willing to provide services in short supply despite the fact that others' unwillingness stems from religious or moral constraints. Thus, the proposed incentives are legally sound.

One other important thing to note about these incentive structures, however, is that licensing boards may be playing a zero-sum game, resulting in a more local version of the brain drain that counseled against adoption of immigration-based tactics. For example, if St. Louis has just the right number of willing physicians, but a county in rural Missouri has none, attempts by that state's licensing board to draw in willing providers could inadvertently create a moral shortage in St. Louis where none existed before. Perhaps this is not terribly problematic, since the redistribution might improve the situation for those in the rural county more than it hurts those dwelling in the city—everyone has some access, even if the level of access is not ideal in either place. Further, the licensing board may be able to recruit additional providers for both locales from an area that has a surplus, potentially from another state. However, even if that is not possible because there is no surplus elsewhere, the licensing board may nonetheless have satisfied its duties with regard to conscientious refusal, as we will see now.

Dealing with True Shortages

In some circumstances it is possible that none of the market incentives aimed at resolving access problems will prove successful simply because a sufficient number of physicians willing and able to perform the service does not exist anywhere; there is no one for the incentives to work on. Here, there is a true shortage rather than a mere maldistribution, and many of the remaining access problems are a result of something other than conscientious refusal.

For example, consider the fact that as of 2007, Washington, D.C., was facing a severe shortage of pediatric rheumatologists due to a confluence of factors including long residencies, relatively low salaries, and bureaucratic problems within hospital practices, but not due to any moral objections to the services falling under the umbrella of that specialty. There are only 144 practicing board-certified pediatric rheumatologists in the entire country, and 30 posts for such physicians are open nationwide. Existing specialists are overwhelmed, and patients simply cannot get in the door, but there is no surplus from which to draw.[80]

If the problem has reached this magnitude in a major metropolitan area, we can assume that things are far worse elsewhere. Nevertheless, resolution will focus on the cities, for there are inherent limits on the supply of physicians able to perform highly specialized medical services that justify setting broad geographic boundaries for access to these services. Even under the best circumstances, there are simply not enough highly specialized physicians available to staff relatively sparsely populated geographic areas. Patients must accept that this is not a result of conscientious refusal per se, but rather is a consequence of cutting-edge expertise taking time to spread, as well as certain services being so specialized that few doctors ever reach that level of skill. For these reasons, patients must be willing to travel farther for ovarian transplants, for example, given their novelty and complexity, than for abortions, which are far less specialized and complex procedures.

Even for services that are not particularly specialized, there may still be shortages of competent physicians, especially in rural areas or during the strain of an epidemic. Regardless of location, it is likely that we have all experienced surprise at just how long it takes to schedule a routine appointment with our primary care physician, and these delays have nothing to do with the physician's moral objections to the sorts of care we are seeking. Rural patients in particular may be used to experiencing delay, having to travel longer distances, and dealing with a new physician even to receive relatively commonplace care; all patients must recognize that there are some inherent trade-offs associated with living far from urban areas and major medical centers. So although it may be desirable for no woman to have to travel more than half an hour to obtain an abortion, for example, the demand for abortions in a half-hour radius of a given geographic area may simply be too low to support a full-time provider, conscientious refusal aside. A full-time provider in this area would exceed demand, and that is not what is required of licensing boards.

Each of these non-conscience-based factors should be taken into account when determining the extent of the licensing board's obligations under consideration here, though there is a need—more like an imperative—to minimize certain shortages and delays stemming from non-moral issues as well. Thus, licensing boards and other interested parties, such as associations charged with accrediting medical school and residency programs, may have an obligation to work toward solutions to remedy this broader divergence of supply and demand, especially by opening the profession to a greater number of applicants and perhaps also by stricter manipulation of the specialty selections of physicians or implementation of some of the options explored above on a larger scale.[81] For the purpose of this analysis, however, remember that we are focused on the responsibilities of licensing boards stemming from their status as the institutional embodiment of a profession that cannot allow personal moral beliefs to be imposed on patients. To avoid such an imposition, licensing boards must simply resolve access problems resulting from conscientious refusal—they need not resolve all health policy issues in one fell swoop.

Nonetheless, it is crucial to recognize that some of the remaining access problems will stem from a shortage of *willing* doctors specifically. These shortages are clearly still a result of conscientious refusals and so licensing boards will need to work to resolve them just as they must resolve maldistributions. However, that will take time to accomplish since it is a matter of actually creating more doctors, not just a matter of resorting those that already exist or even excluding refusers, which is undesirable for reasons already described and, as we will see in the next chapter, would not necessarily even resolve shortage problems.

Boards should absolutely be expected to make progress in this area, but because they lack any immediate options, the geographic boundaries in which boards are responsible for providing access to willing physicians ought to be drawn with constant consideration for the number of physicians currently available for recruitment. These boundaries must be continually readjusted as board interventions successfully generate more willing physicians, but there is no overnight solution, so patients may still be expected to travel farther for services that really ought to be available more locally. Nevertheless, this incremental approach is the best option and is preferable to the existing state of affairs.

Even with this limitation of their responsibilities, however, licensing boards need guidance as to how many willing physicians in a particular geographic area will suffice, given that shortages are only shortages by comparison.[82] To address supply-demand discrepancies resulting narrowly from conflicts of conscience and to truly define the term adequate so that we can assess whether licensing boards are fulfilling their obligations, the use of benchmarking methodologies will be quite useful. A board can select a geographic area that it believes has achieved acceptable availability levels relative to the overall supply of available physicians, even if these levels are less than ideal, and use that area as a measure of whether demand is being appropriately met somewhere else. Importantly, benchmarking provides a "guidepost that does not depend on a hypothetical optimal physician level but [rather] depends on a real-world and attainable health care system."[83]

Of course, licensing boards must take care to choose appropriate benchmark areas. They must be sufficiently similar to the geographic region in question with regard to relevant characteristics, such as patient attributes, and it will be important to consider baselines in different parts of the country, noting the type of delay patients normally experience when seeking various types of care. Further, boards must develop standards that can be used to measure whether demand is being adequately met in the benchmark area. Existing managed care standards for staffing geographic areas may be helpful, and licensing boards might also consider whether the levels of physician deployment witnessed in potential benchmark regions suffice to avoid a significant loss in patient welfare.[84] Finally, boards could draw from other measures of health-care system efficiency, such as the World Health Organization's Health Systems Performance Assessment, which evaluates how well resources are being used to meet socially desirable health-care goals.[85]

While certainly not an exact science, these considerations should assist licensing boards, and potentially courts if the licensing board's choices and initiatives are challenged as deficient, when determining whether patient demand has been satisfactorily accommodated despite conscientious refusal. The essential thing to take away is simply that where there is a true shortage, the geographic areas in which a licensing board has an obligation to make a particular service available to patients must simply be drawn more broadly, at least for the time being.

Because merely imposing the burden without providing more detailed guidance would be insufficient, the preceding chapters have developed the proposal that licensing boards, as the most appropriate representative of the medical profession as a collective whole, should bear the responsibility of resolving access problems created by the conscientious refusals of individual physicians. Licensing boards must determine which demands they have an obligation to satisfy by ensuring access to willing physicians, measure patient demand for the identified services, establish a registration system for refusing physicians that will simultaneously enable them to measure the supply of willing physicians and to reject physician refusals that are based on inappropriate grounds, determine appropriate geographic boundaries in which the supply of willing physicians must be capable of satisfying patient demand, and most importantly, take measures to remedy mismatches of supply and demand resulting from the conscientious refusals of physicians in a given area.

While certainly a challenge, fulfilling these tasks will allow licensing boards to overcome, or at least mitigate, the market failures that currently make conflicts of conscience between doctors and their patients so problematic. Their remedial actions will improve patient access and choice, while refusers will stay afloat only if there is a sufficient number of patients willing to choose them in light of the available alternatives, but they will at least have a chance to practice in accord with their personal moral beliefs. These are important goals, and if we consider a compromise sufficiently worthy of accomplishment, as we should, the licensing board solution is the most promising way of bringing it about.

8

The "Hard" Cases: When the Institutional Solution Fails

We have now seen what is expected of the licensing board, but what if it fails to remedy patient access problems in the way that is expected of it because the board is not appropriately working to resolve shortages or maldistribution of willing physicians, or it simply has not yet been successful in doing so? Patients will be left without reasonable access to services that we have determined they have an interest in receiving, if not a right, and we are left to decide among only second-best solutions, namely, (1) eliminating the option for conscientious refusal by physicians in these situations, such that they must either provide the requested service or face disciplinary or liability consequences, or (2) maintaining the permissibility of conscientious refusal, while offering patients recourse against licensing boards.

At the outset, it is essential to separate these "hard" cases, categorized as such because either patients are left without care or physicians are forced to act against their fundamental beliefs, from those situations in which supply and demand can be—and are—calibrated. The mere fact that the hard cases can sometimes exist is no reason to demand at all times that physicians violate their moral convictions regardless of the alternatives actually available. The exception should not drive the rule and context does matter. In fact, before health-care providers can be expected to undertake a given risk, health-care institutions have an obligation to minimize that risk to the fullest extent possible. Thus, as we saw previously, before demanding that physicians provide care to SARS patients, for example, hospitals, or even the government, must properly equip those physicians with the best means available to minimize their risk of infection.[1] The situation in which no such equipment is available is much different from one where it is, and should be treated accordingly.

Similarly, if licensing boards can minimize the scenarios in which physicians will be confronted with conflicts of conscience, they must do so. Because many of today's access problems appear to be a result of a maldistribution of physicians,[2] appropriate incentives offered by licensing boards, as well as efforts to encourage moral diversity in the profession, should successfully mitigate, if not eliminate, many of the problems otherwise associated with conscientious refusals. In other words, the protective equipment is often available; redistribution is key. In fact, even if a genuine shortage of physicians remains as a result of conscientious refusals or other factors,[3] that would shift the appropriate benchmarks; fewer willing physicians would have to suffice for the time being, even if their numbers do not truly meet patient demand.

Clearly, in these latter situations, the conflicts associated with hard cases will remain at least for a time, as they will when boards simply fail to properly redistribute existing willing physicians. However, for a variety of reasons explored in this chapter, the institutional solution should not be abandoned, even under these more complicated circumstances. Instead, patients should look to licensing boards for compensation and continued efforts to avoid the hard cases.

Justifications for Prohibiting Physician Refusal in "Hard" Cases

The inability of patients to obtain a variety of medical services from non-physicians, the threat of informal rule by an oligarchy of physicians, and the value of diversity in terms of the medical services that are available are all factors suggesting that the last doctor in town should be expected to provide the requested service regardless of his personal moral objections.[4] Notably, the imposition of such an obligation, or at least the elimination of protection for refusal in the hard case, would not be a function of the patient's autonomy trumping the physician's per se, which was Pellegrino's major concern. Instead, the physician's moral autonomy is trumped by her own autonomous choice to enter a profession that bears social benefits and social responsibilities. Personal integrity is worthy of protection, but it can be outweighed by the need for professional integrity.[5]

As we have seen, the physician voluntarily accepted the role of a professional, presumably with at least the implied understanding that the profession as a whole has obligations to make services available to patients vulnerable to its collective monopoly and that circumstances may arise in which the physician will stand as the sole gatekeeper, thus bearing the profession's normally collective obligations. Even if physicians have not been asked to consent to these broader responsibilities as a result of the expansive permission for refusals currently offered by statements of professional ethics and the protection extended by existing conscience clauses, both of which tend to be granted without regard to whether the refuser is the last doctor in town, doctors should be expected to consent now. They need not always subordinate their conscience to patient demands, but must do so when the patient has no other choice.[6]

Doctors have the flexibility to control their exposure to situations they find morally problematic by virtue of their selection of employment and geographic location, and they may avoid having to consent to moral risk if they take action to ensure that they will never stand as the sole gatekeeper for any patient. However, if they have failed to do so, their voluntary choice renders their interests less compelling than those of patients depending on them for care.[7] It is precisely this voluntariness—if not outright consent to responsibility—juxtaposed against the less autonomous limitations on the patient's freedom established by professional monopolies that could justify forcing physicians to set aside their personal moral qualms in the hard cases or face the attendant consequences. This existence of choice also differentiates conscientious refusal in medicine from conscientious objection to war and exemption from the mandatory draft. Like attorneys, physicians occupy a state-licensed gatekeeping role and have "deliberately chosen a path of education and licensing expressly devoted to the performance of a quasi-public function, empowering [them] to facilitate access to a system of rights and privileges that is integral to the well-being of every citizen."[8]

Unfortunately, some commentators have inappropriately ignored this voluntariness differential. For example, Karen E. Adams allows a balancing between the physician's and patient's interests even in the hard cases, and would permit a physician to refuse care when provision of the service would damage the physician's integrity more than denial of the service would harm the patient.[9] Not only would such a comparison be incredibly subjective—how does one contrast the harms at stake?— but this balancing of interests, as though both doctor and patient had equal choice in the situation they face, disregards fundamental differences between them. Of course, one might also argue that patients have made a voluntary choice to reside in locations where they are likely to have difficulty accessing the medical services they desire, and we saw previously that such patients will often have to accept reduced access and make sacrifices of their own. The choice of where to live, however, may be far less autonomous than the choice to become a physician. It may be influenced by a variety of factors outside of the individual patient's control, such as financial constraints, educational limitations, and responsibilities and connections to family members; this is precisely why we do not see a mass exodus from areas with terrible public education systems, high crime rates, or generally poor politically leadership. Further, in some contexts, a patient may truly not have made a choice at all, such as the teenage girl who is fully dependent on her parents but is seeking contraceptives or an abortion without their knowledge or consent, or a traveler facing an emergency situation. Most importantly, the patient's choice of location, where it exists at all, does not involve accepting a duty owed to others, which is precisely what entry into the profession of medicine entails. Thus, it is not the case that voluntariness on the part of both doctor and patient cancel one another out.

Fenton and Lomasky point out other relevant differences between professionals and their patients, explaining in the context of pharmacist refusals that "when individuals confront one another as moral equals, they are not (barring exceptional circumstances) obliged to render more than simple noninterference with the projects of others. The crux of our argument[, however,] is that *in the case under examination, moral equality does not obtain.* That is because the pharmacist is in a privileged position vis-à-vis potential clients. For a wide range of drugs, only physicians are legally entitled to prescribe, and only pharmacists to dispense. An individual who wishes to secure one of these is limited in her choices to those designated channels of supply. Her freedom to cooperate with willing others is thereby limited."[10]

They go on to note that health-care professionals with monopolies over the services they provide have advantages that have been secured to some extent at the expense of patients, although at least implicitly with patient consent through the social contract permitting such monopolies in the first place.[11] The restriction on patient liberty is not wrongful per se, since the reason such professional monopolies exist is due to a social judgment that limiting the practice of medicine to physicians is in the best interest of the public's health. However, the fact that the restriction exists at all allows what would otherwise be very strong claims on the part of refusing physicians to be defeated in the hard case.

When the refusing physician is actually or effectively the last doctor in town capable of satisfying the patient's request—the only person who could open the gate—the responsibilities of the profession of medicine as a whole, the very basis for placing responsibilities on state licensing boards, can no longer be separated from individual physicians. Instead, they trickle down to the individual physician who now personally bears the monopoly, and thus bears the heightened obligation of providing access to the service the patient desires, even though he would have no such obligation in another context. With this professional burden, it is no longer appropriate for the physician to exercise conscientious refusal based on his personal moral beliefs.

Note that this is exactly the conclusion reached by Robert Goodin via his principle of individual responsibility—the more others depend exclusively on you, the greater your responsibility to assist them.[12] Applying his principle, Goodin argues that "a professional ought to be held morally and professionally (if not legally) responsible to provide assistance in any situation in which, if he did not, someone would be left helpless." Hypothesizing about a lone doctor in an isolated frontier town, the quintessential hard case, he correctly notes that the doctor is not morally entitled to withdraw from the case of a patient too ill to be taken elsewhere, regardless of the provision of notice that he plans to do so, nor can he refuse to take on the patient in the first place. Neither move would eliminate the patient's unique vulnerability to him.[13]

Similarly, at least one court has recognized that patient access is a primary determinant of physician duties, and restrictions on access can serve as the basis for restricting widely accepted professional freedoms. In *Leach v. Drummond Medical Group*, the sole physician practice group in the area refused to accept a patient who had filed a complaint against one of the group's doctors. The court, however, found a common law duty to treat, despite the fact that physicians are ordinarily allowed to refuse patients on any nondiscriminatory grounds.[14]

For all of these reasons, the general principle allowing individuals to reject coerced cooperation does not seem to hold in hard-case settings, but a few important questions remain. First, we must resolve the lack of clarity regarding areas facing only an artificial shortage as a result of several physicians registering refusals to provide a given service—which of those registered refusers would really stand as the last doctor in town with obligations to the patient? Is it the physician who was unlucky enough to have the patient seek care from her first, or is it the last one asked after a string of other doctors have refused? It appears that each would-be refuser in areas where supply fails to meet demand should bear equal responsibility, but clearly the patient only needs one doctor to act.

In this situation, a lottery system may overcome the potential unfairness of enforcing an obligation held by all against only some and exempting others. For example, James Childress has suggested that if the state could not get enough people to serve in the military because too many were conscientiously objecting, the fairest way to deal with the problem would be through a lottery to determine who would be forced to provide the service despite their moral opposition.[15] The scenario is almost exactly the same for physicians if we conceptualize the randomness of a patient's selection of any given physician as the equivalent of a lottery. However, the patient's selection is not likely to be entirely random and a particular physician may face more than his fair share of requests for objectionable services simply by virtue of his office being in a convenient location or his name being listed first in the phone book. This potential for unfairness could be overcome by a true lottery or a system analogous to the court appointment of attorneys, with licensing boards selecting some area physician to provide the requested service.

With that issue disposed of, we must also contend with the argument that physicians with moral objections to a patient's request might legitimately refuse even if the patient lacks alternatives because whether or not there is another willing provider does not change the fact that there is never an obligation to participate in genuine wrongdoing. It is certainly the case that no professional obligation could rightfully include a duty to engage in true moral transgressions, regardless of voluntary entry into a profession or the existence of a professional monopoly. However, it is essential to recognize that allowing individual physicians to personally determine which medical services fall under the heading of "wrongdoing" would permit the physician's conscience to become a law unto itself in these hard cases, imposing on patients views with which they may reasonably disagree.[16]

Conscientious action can often be described as apolitical, given that it does not generally result in complete cessation of the activity that the actor wishes to avoid. This would be the case when a physician refuses a patient's request on moral grounds, but the patient is able to obtain the service with only mild inconvenience from another physician. Individual political action, on the other hand, can have the consequence of terminating the activity in question provided the actor has sufficient power, and this precisely describes a physician's refusal in the hard case—if she will not provide the abortion, the patient will not obtain one, not legally at least. When individual action has this effect, it is an illegitimate use of power because the objector, purposely or not, usurps the authority of government bodies (here, either the legislature or state licensing board) that have either legalized or failed to prohibit the medical service in question, thereby accepting it as a legitimate patient demand.[17]

The proper place for physicians to deal with these issues is at the level of shaping health policy in a democratic fashion, not through misuse of the collective professional monopoly.[18] If they fail to convince others of the genuine moral wrongness of a particular service that is now being requested of them, they can still say no, but they have lost the debate as to the profession's responsibilities and can no longer stand in as an appropriate representative of that profession as a whole. Thus, their refusal as the last doctor in town would be unacceptable.

On the other hand, there are some conscientious refusals that may in fact be obstructive but that are nevertheless perfectly appropriate,[19] and potentially obligatory. For example, even the last doctor in town only has an obligation to satisfy patient requests that fall under the demands the licensing board determined it has a responsibility to meet. Martha Swartz reaches a similar conclusion, arguing that we must distinguish between *"professional* integrity based on prevailing medical ethics" and *"personal* morality." She encourages health-care professionals to refuse to provide services that would violate "generally accepted professional standards of practice," while precluding them from "injecting their personal beliefs into their professional practices."[20] This professional/personal distinction has also been made by several cases.[21]

Unfortunately, Swartz does not give enough weight to the personal morality of physicians, since she fails to recognize a difference between the heightened obligations present in the hard cases and the greater freedom to refuse when other willing providers are available. More importantly, she fails to address the potential insularity of professional ethics, allowing the profession to set its own standards for obstruction based on its collective conscience involving moral—not purely medical—judgments without necessarily involving other interested stakeholders. Without such broader participation, these standards of professional ethics represent nothing more than a different species of conscientious objection capable of imposing debatable perspectives on others, and therefore, must be treated as such.[22] In other words, even refusals based on prevailing medical ethics may be impermissible in the hard-case scenario when those moral standards have not been appropriately vetted through the mechanisms described previously.

However, just as licensing boards can reject demands for scientifically invalid services, physicians in the hard cases could also permissibly exercise refusal on grounds of medical expertise given that some services within the licensing board's purview may not be clinically indicated for particular patients. Even the last doctor in town avoids inappropriately interfering with patient autonomy when he relies on reasons for refusal that are actually part of the profession's special knowledge base, the very reason it has gatekeeping power in the first place.[23] It is important to recall, however, that many decisions made in the health-care context are not in fact based on this special knowledge, such as decisions about futility, benefit, resource allocation, or other value-laden choices.

It is possible that a licensing board given more freedom to restrict its obligations beyond the simple legality/scientific expertise standard will define these difficult terms and potentially exclude certain things from the repertoire of services it must make available, thus allowing physicians to refuse to provide those excluded services in all circumstances.[24] However, if licensing boards have failed to make these value judgments, or were never given the authority to do so, these decisions must remain the patient's. Allowing physicians to make determinations in the hard case about which services are optimal or "suitable" is problematic for patient autonomy and informed consent, since as we saw previously, only patients themselves have the right to decide which benefits are personally worthwhile and which risks are too great to bear.[25] Thus, a physician's refusal to provide services he believes to be suboptimal (but that are still clinically effective for the patient's aims) is a value judgment that cannot be appropriately exercised when the patient lacks access to any alternative provider.

Aside from these very limited situations, physicians finding themselves representing the professional monopoly all on their own have a professional obligation to provide requested services to patients regardless of their own moral beliefs. No one has the right to impose his personal ethical values on anyone else,[26] and whether that is their intention or not, it is the effect of physician refusals in hard-case scenarios. Thus, refusing physicians can *legitimately* be forced to accept the consequences of their disobedience as compensation to the community, which is being forced to bear the costs of the dissenter's evasion of duty.

Nevertheless, for many of the same reasons that courts will not specifically enforce employment agreements—such as the problems associated with guaranteeing quality performance, the undesirability of forcing continued personal relationships after confidence has been destroyed, and, most importantly, the distasteful resemblance to involuntary servitude[27]—physicians should not actually be compelled to provide services they find morally objectionable. There is also the larger problem associated with the impossibility of physically forcing another person to do something he does not want to do, and we cannot ignore the fact that most patients would likely not want care from a doctor compelled to provide it. As stated bluntly by a representative of the National Women's Law Center, "I don't want a doctor performing my abortion if he has moral reservations about it. Thanks but no thanks.... It's not in the client's interest to have [his attorney] despise him and represent him, just like it's not in [the patient's] interest to have a doctor who is shaky or morally ambivalent performing [her] abortion."[28]

Even so, there are significant and tenable reasons to at least demand that the last doctor in town choose between providing a service and facing professional discipline and/or liability. The distinction between proper and improper grounds for refusal in the hard cases elaborated above might even serve as malpractice standards of care in suits brought by patients against refusing physicians. The costs associated with such consequences will certainly serve as significant disincentives to refusal—but they may also serve as disincentives against becoming a physician or specialist entirely if unwilling to handle the obligations associated with hard cases. It is precisely the extent of these disincentives, and their unintended consequences, that make the imposition of penalties for hard-case refusals too troublesome to accept. Despite their significant justifications described in this section, they may simply do more harm than good.

Problematic Disincentives

If physicians had an enforceable legal obligation to meet patient demands in the hard cases regardless of their own moral qualms, they might also have an obligation to receive training in and maintain the competence to provide all services for which licensing boards bear a responsibility within their given medical specialty. Why would this be the case? Because in order to count as the last doctor in town, a doctor must be able to perform the service in question. Thus, a psychiatrist will have no obligation to a woman seeking a tubal ligation because to have a duty to the patient, the physician first must have the technical skills needed to fulfill it; this physician's services are obviously not a relevant option for this patient.

For this reason, the conscientious refuser could simply avoid any hard-case duty he finds morally objectionable by purposefully avoiding the requisite clinical skills. Unwilling providers would essentially be disguised as incapable providers, exercising at an earlier point in time the conscientious refusal that would otherwise be impermissible when they stand as the last doctor in town. Clearly, however, the access problems associated with such refusals would remain. Enforcing hard-case obligations without imposing a supplemental competency requirement would have little practical effect for some services, since it is likely that few physicians already possessing the technical skill required to spark these obligations would need the threat of punishment to fulfill them. Therefore, all doctors might be expected to at least prepare themselves adequately to provide the care requested by patients coming to them as a last resort, though of course the psychiatrist need not learn to perform reproductive services and the usual specialization of expertise would still be acceptable.

Notably, the AMA seems to recognize a similar duty, at least in part. Despite the association's broad permission allowing physicians to refuse to provide services that violate their personal conscience, the AMA has stated with regard to medical students that "medical schools should have mechanisms in place that permit students to be excused from activities that violate the students' religious or ethical beliefs *[but] ... students should be required to learn the basic content or principles underlying procedures or activities that they exempt....* Patient care should not be compromised in permitting students to be excused from participating in a given activity."[29]

However, requiring ob-gyns morally opposed to abortion, for example, to not only learn how to perform that procedure, but also to perform or assist in the number of abortions necessary to sustain continued competence, would eliminate a great deal of the compromise fashioned by the institutional solution, since doctors would have to provide at least some abortions even when they were not the last doctor in town. It is quite likely that many physicians who could not bring themselves to perform this procedure under any circumstance, let alone in mere preparation for the possible hard case, would simply choose to enter other specialties in the face of this competency requirement.[30]

In other situations, the physician may already be clinically competent to provide the requested service without the imposition of any supplemental training duties, such as the doctor who is morally opposed to terminating life-sustaining care but who is nevertheless capable of removing a patient's feeding tube or the physician who was previously willing to perform abortions but has undergone a religious conversion. These physicians might predictably accept the disciplinary or liability consequences of refusal in the hard case as an alternative to actually performing the objectionable service, or might even leave the geographic area, specialty, or profession of medicine in order to avoid hard-case scenarios altogether.

Although many commentators seem to think that this would be for the best, this is precisely the sort of exclusion of moral refusers that we must avoid if we are to retain the counsel of restraint and many benefits of moral diversity in medicine. Further, such unwilling individuals may not be immediately replaced by those who are willing to provide the services in question, if licensing boards are able to attract or generate more willing physicians at all. Additionally, if potential refusers avoid a particular field entirely, we may be left with a shortage of doctors in that field, but it may not be clear that the shortage is a result of conscientious refusal. Thus, it may slip through the cracks as not evidently part of the licensing board's obligation to resolve.

Clearly, patients in the hard cases may not actually be helped by policies that punish the last doctor in town for the breach of his professional duty. In fact, patients would very likely be in precisely the same situation that they would have been had that doctor originally been permitted to refuse, for they would still not receive the desired service when and where they should. Moreover, the expected physician responses could exacerbate existing availability problems. If the refuser was the sole ob-gyn in the relevant geographic area and she decides to relocate or quit rather than perform abortions, patients seeking morally uncontroversial services may also be left without a provider, thereby indirectly, but severely, affected by elimination of conscience protection in the hard case.

To meet competing interests, not the least of which is the protection of physicians, the state should have the obligation to show that its interests are sufficiently important and unable to be realized unless protection of physician conscience is overridden in a particular circumstance.[31] Patients' interest in the availability of services is unquestionably important, especially given the likely centrality of availability to the social agreement allowing the professional medical monopoly in the first place. However, it is simply not the case that access can be realized only through denial of conscience protection to the doctor bearing the professional monopoly all on his own. To the contrary, enforcing a physician's hard-case obligations at the point when he is actually facing the hard case may not improve access for particular patients and may actually restrict it for others, while enforcing them at some earlier point would destroy any middle ground capable of balancing the interests of doctors and patients. Thus, not only is physician coercion incapable of solving the problem, but it also sacrifices much of the value that physician refusers add.

Whether our primary concern is for doctors or patients, as an enforcement matter we would be well advised not to distinguish the easy cases, in which there exists a supply of accessible, willing physicians, from the hard cases, in which no physicians willing to satisfy the patient's request exist within a reasonable geographic range. Doctors in both cases should be legally permitted to refuse.[32] In fact, while physicians should be strongly encouraged to respond to patient demands when the patient has no practicable alternative, and have a genuine professional duty to do so despite their own personal moral objections, punishment for breaching this duty should be limited to nothing more than social approbation. Even if a patient has been entirely denied access to some service within the board's purview as a result of conscientious refusal, he should not have a cause of action against any physician. The sting of this conclusion is heavily mitigated, however, by the baseline conditions for the exercise of a physician's conscientious refusal described subsequently, particularly the enforceable obligation to provide care in emergency contexts. The refused patient will generally also have some alternative recourse even in nonemergency hard-case scenarios.

Licensing Board Accountability

In light of the many problems associated with extracting a penalty for a physician's hard-case refusal, retaining an institutional approach is the most prudent and effective way to address these situations. This conclusion is strengthened by the fact that, as we just saw, patients seeking services from refusers are unlikely to receive the service anyway, because refusers will often choose to face the consequences rather than perform the service. Considering that monetary compensation will be their only remedy, patients will not suffer any greater harm if they are compensated by licensing boards rather than by the refusers themselves. Further, the institutional solution paired with institutional penalties places incentives for change on the body most capable of implementing the mechanisms of such change. Thus, when the patient is denied satisfaction of his demand for a service that the board has an obligation to fulfill in the area in which the board has an obligation to fulfill it, the patient should be able to hold the licensing board accountable for its failure.

If that is the case, how should accountability be enforced? Surely, holding the elected officials who appoint the licensing boards responsible at the ballot box would be too attenuated. It might also be ineffective, since majoritarian impulses, especially in certain states, could overwhelm the protection of access to services that have been determined to be part of the board's charge. Even if some women had to travel much farther for abortion services than they should have as a result of the board's failure, for example, it is quite possible that those women would be unable to raise sufficient concerns among the majority of voters to sway an election. In fact, this majoritarian difficulty will be one of the biggest problems in convincing the states to adopt the compromise proposed in this book. However, as more types of refusals affecting more types of patients begin to arise, a greater proportion of the population will become concerned about accessing services important to them and the call to action will grow.

Clearly, more direct accountability is required for licensing board failures, likely through a claim against the board itself resulting in a monetary award for the aggrieved patient. Notably, these claims should turn out to be relatively infrequent, considering not only that boards have the tools available to redistribute willing physicians to shortage areas, but also that the geographic boundaries in which access must be preserved and the definition of reasonable access itself are based on the supply of willing and able physicians currently available for recruitment. In other words, boards are not being asked to accomplish any impossible task, and if a particular obligation appears impossible to fulfill, it is in effect a sign that geographic boundaries have been drawn too narrowly with regard to a particular service. However, there will be occasions where licensing boards fail to meet their obligations to deal with conscientious refusals even as limited by these practical caveats. When that occurs, the question becomes how stringent the board's accountability to patients ought to be.

Under a negligence standard, the unserved patient would have to show that there were reasonable steps to ensure access available to the board that it nevertheless chose not to take. However, patients would have to demonstrate that the unused alternatives would have been successful in avoiding access problems, and they would also have to contend with public policy defenses, such as competing demands on public resources. Further, application of the reasonableness criterion could presumably result in some measures being deemed unreasonable and boards could thus avoid having to implement them, creating some circumstances, even if very few, where patients whose demands should have been satisfied in the relevant area but were not are left without compensation from any source.

Before determining whether this is acceptable, however, it is crucial to recognize that as a result of the way the relevant boundaries defining board obligations will be drawn, we have already accepted a similar situation, at least temporarily, in which some patients who have been denied appropriate access as a result of conscientious refusal will not be compensated by either refusing doctors or licensing boards. For example, the board may be in the process of resolving a true conscience-induced shortage, but so long as it is satisfactorily ensuring access in the relevant area—the area that is bigger than it truly ought to be because it takes account of that shortage—it has not failed to fulfill its current obligations. In that context, the patient would have no claim against the board, even though without conscientious refusals, she would have had access to the desired service in a tighter geographic area than that in which she was actually able to obtain it.[33]

Certainly, the board should not be penalized while its hands are tied and it can do no more at the moment to improve matters, but the fact remains that there will be a temporary gap in full protection of patients' interests. The same is true while the board is working to fulfill its existing obligations in the relevant geographic area, for it must be given some reasonable period of time in which to implement the mechanisms described in the previous chapter before accountability will attach. The licensing board approach does not promise to instantaneously resolve the problems associated with conscientious refusal, however, and it is not the case that patients will be compensated whenever their access to desired services has been inappropriately burdened on these grounds, although that is the eventual goal. While this is a rejection of true strict liability for licensing boards, it leaves patients no worse off than they are under the current conscience clause system and it offers the prospect of marked improvement down the road.[34]

The issue that remains is whether the possibility for this no-compensation outcome should be expanded with regard to area shortages that boards currently have the capacity (and thus, the obligation) to remedy, but that remain even after they have been given a reasonable opportunity to do so. This may be acceptable if we think that there truly are situations in which it would require an inappropriate dedication of resources to successfully ensure the availability of a service otherwise falling within the board's obligation. On the other hand, if it is sufficiently important to minimize to the fullest extent possible those situations in which patients have to bear the burden of a physician's personal moral beliefs, holding boards accountable whenever they fail to ensure patient access in the relevant geographic area may be more appropriate.

The ideal method of dealing with board failures, and one capable of splitting the difference between these two alternatives, will be closest to a regulatory system of fines. This is the preferable punishment in other areas of conscientious refusal, such as the pharmacy context,[35] since it does not depend on the willingness or motivation of patients to bring suit. Of course, such a system could not quite work here because imposing fines on a government agency would simply be a case of transferring money from one pocket to the other—clearly, the "fines" must be paid to some outside entity for their impact to really be felt.

The next best thing is for boards to be expected to compensate the patient whenever she can show that she went unserved for a covered service in the relevant area. The board could avoid having to pay compensation only by showing that there was indeed some willing provider geographically available—that is, a physician within the designated geographic area who at the very least had appointments available. Note, however, that given the highly limited nature of our inquiry focusing on the board's obligations to deal with conflicts of conscience specifically, financial accessibility of that provider to that patient is not necessarily required, even if this is also a normatively appropriate goal.

Injury to the patient will be presumed from the simple denial of services that the licensing board had a responsibility to make available. No demonstration of physical harm is necessary, considering that licensing boards will have an obligation to preserve the availability of even elective medical services. The patient also need not prove that she went unserved as a result of conscientious refusals specifically, since that is accounted for in the geographic boundary itself. Compensation could feasibly be obtained by filing a claim with the board, implementing an efficient administrative approach subject to judicial review similar to that used for Social Security disability benefits, for example, or the unserved patient might alternatively seek compensation by going directly to court in the first instance.[36]

The applicable compensation schedule could be set by statute, or perhaps by notice-and-comment rulemaking, but either way, larger awards ought to be made for the denial of services more critical to a patient's well-being. Further, awards would have to be appropriately calibrated to simultaneously accomplish two things. First, they must avoid board incentives to shirk obligations to ensure actual access to important medical services where it would simply be cheaper to pay compensation. However, they must also permit boards to behave efficiently with regard to less important services, allowing them to pay compensation when it would be unreasonably expensive to satisfy patient demand.[37] This differentiation of awards avoids imposing unjustifiable demands on licensing boards, while also working toward the goal of ensuring that patients either receive the services they desire or are compensated for violation of the board's responsibility.

This payment system appropriately favors patients, and essentially ensures that the "fines" are paid to the right people, namely, those whom licensing boards have failed. However, because a licensing board's defense would be limited to actual availability, the drawing of geographic boundaries for relevant access areas takes on added significance. Given how crucial this determination is to board accountability, leaving it up to the boards themselves appears to introduce a grave conflict of interest. For example, there may be some concern that if licensing boards bear the responsibility of defining the services for which they must ensure patient access, as well as defining the geographic area in which the service must be made available, they could simply avoid accountability by eliminating from their list services for which they are unable to meet demand, expanding the geographic range unreasonably, or revising the benchmarks used for comparison.

Clearly, this sort of gaming behavior would be professionally irresponsible and inappropriate for licensing boards to engage in, but hoping that they will just avoid it on their own is naive. Perhaps it is necessary to have a separate agency draw these limits and obligations, one that is not simultaneously responsible for satisfying them. However, this would be quite inefficient, since the legislature would still have to define the delegation of authority to some other body, and this separate body would still be an instrument of the state interested in avoiding liability for the state. Alternatively, the legislature could simply avoid delegation and set the precise limits of the board's obligations itself, but this solution is untenable because legislatures simply do not have the institutional resources to make these very detailed decisions as carefully and as specifically as they must. The legislature should certainly provide standards to govern board behavior and decisions in this area, and we will see an example of these standards in the final chapter, but delegation offers important benefits that ought to be utilized. Fortunately, the problems associated with conflict of interest can be managed.

First, as discussed previously, boards ought to use a system of notice-and-comment rulemaking to draw (and update) the relevant geographic boundaries and other lines defining their obligations, and with notice-and-comment procedures comes transparency and involvement of the public. Perhaps more importantly, licensing board actions will be subject to judicial review, allowing courts to step in to ensure that boards have set appropriate areas in which services must be available, have not excluded medical services from their responsibility that should fall within it, are appropriately measuring patient demand and physician supply, and are taking logical steps to resolve maldistributions and true shortages stemming from conscientious refusal.

Patients should not be able to recover against licensing boards just because they could not access a service they wanted in the area where they believe it should be available, but they should be permitted to challenge a board's decisions and behavior. The standards for judicial review of administrative actions, in this case the actions of licensing boards, are well recognized and are in fact quite well suited for the institutional solution proposed here. The decisions confronting licensing boards will often be quite difficult and they will usually have a variety of permissible options from which to choose, each with its own benefits and drawbacks, and each that might be subjectively preferred by outside observers. Additionally, courts would have no special expertise to make these decisions themselves. Therefore, judicial review will be limited to making sure that licensing boards have adopted an approach that is at least within a range of permissibility—that they have followed appropriate procedures and have acted in accordance with the governing statute. Boards will be required to behave reasonably in satisfying their obligations; courts will overturn only those board actions that are arbitrary and capricious.

Licensing boards should certainly pay compensation when they fail to satisfy their own standards, but even when they have succeeded, those standards themselves are open to challenge. The important thing to note is simply that patients will not be subject to the unchecked whim of licensing boards. In most cases, patients will be compensated when denied access as a result of conscientious refusal and licensing boards will have an incentive to continually work to avoid that very problem, as well as the discretion to determine the best ways to go about doing so.

The hard-case scenario has been the fundamental driver of the conscience clause debate, with reasonable commentators on all sides focused on the essential question of what to do when a patient's only practicable option for access to a desired medical service is a physician who is morally opposed to providing it. Eliminating these scenarios altogether is the most promising path to a compromise, and licensing boards have the capacity to work toward that goal. However, when they fail, or simply have not yet succeeded, it may seem like all is lost and we are once again forced between choosing to protect the refusing physician or the vulnerable patient. In fact, although physicians may have a moral "right to object, they should not have a right to obstruct,"[38] and the last doctor in town really should comply with the patient's request, given that he now bears the profession's monopoly power and concomitant obligation to preserve patient access.

Unfortunately, enforcing such an obligation against the physician would weaken the institutional solution and would provide little additional protection for patients. It is quite likely that objecting physicians would choose disciplinary or liability consequences rather than violate their deeply held moral beliefs, that patients would generally be wary of placing themselves in the hands of physicians opposed to their requests, and that the existence of serious disincentives would drive otherwise good doctors out of various specialties or away from medicine entirely. Holding licensing boards accountable when they fail to meet their obligations is a superior solution, one that maintains the value of an institutional compromise from top to bottom, even if access is still sometimes denied and compensation is not currently made available to every single patient for whom access is inappropriately burdened as a result of conscientious refusal.

9

Physician Obligations and Sacrifices

So far, we have explored the ways that licensing boards can offer a promising, albeit imperfect, solution to the conscience clause dilemma currently facing the medical profession. But while these boards will clearly bear a great deal of responsibility, physicians themselves must bear some burdens of their own in order to be free of liability to patients or other consequences stemming from the exercise of conscientious refusal. There are some professional duties that are not only collective, but rather truly are applicable to—and ought to be enforceable against —each individual physician.

In addition to refraining from refusals based on invidiously discriminatory patient characteristics, physicians should only be granted conscience clause protection if they, at the very least, fulfill their obligations to promptly notify patients of the services they will not provide on moral grounds, offer patients information regarding all of their medical options even if the physician is unwilling to perform some of those options, and provide care to the best of their ability to patients facing emergency situations. Many commentators would also add referral obligations to this list, but we will see that the imposition of this sort of duty is somewhat more difficult to justify. Further, while physicians should be free to advise patients as to what they feel is the right thing to do, they must avoid religious coercion, proselytism, and moral arrogance, since these "are abuses of the power entrusted to the physician's noble vocation."[1]

While these baseline responsibilities are likely to be relatively uncontroversial on the whole, given that they require the provision of the actual service that the physician finds objectionable only in very narrow and infrequent circumstances, enforcing these obligations could potentially give rise to some of the negative disincentive effects that just counseled against imposing liability or other consequences for refusal in the hard case. Nevertheless, these responsibilities are so basic and so essential to the autonomy and well-being of patients that we simply cannot afford to sacrifice them. If physicians desire protection of their conscientious beliefs in a professional setting, this is the price they must be willing to pay.

Before addressing these nonnegotiable duties in more detail, however, it is essential to point out that even these can often be mitigated by the behavior of patients, and it is not at all outrageous to ask them to make relatively minor sacrifices in order to respect physician conscience. In this regard, Veatch notes that patients must show consideration for the autonomy of their physicians,[2] and Pellegrino argues that the "physician-patient relationship is a moral equation with rights and obligations on both sides ... it must be balanced so that physicians and patients act beneficently toward each other while respecting each other's autonomy."[3] Along similar lines, the AMA states that patients must work with their physicians in a "mutually respectful alliance."[4] Although this statement almost certainly was not made with the concept of conscientious refusal in mind, it might be interpreted to require some consideration of the physician's morality before a patient demands services from him or her that the patient could obtain with relative ease from another source.

Just as patients have a responsibility to bear in mind the claims of the wider community when they make demands for care, their obligation to consider the impact of their choices on others can be extended beyond the context of resource allocation to fair consideration of their physician's interests.[5] If one takes the idea of moral risk seriously, we can draw an analogy to patient obligations to physicians during epidemics. In that context, Sokol argues that the virtue of tolerance demands that patients

acknowledge healthcare workers' plurality of roles, as well as their fears and concerns in the face of severe risk. If these fears are well founded and reach such a level that medical staff are worried for their life or that of their loved ones, the virtuous patient ought to allow them to step down from their role as caregivers. In such cases, insisting that they continue in this role would reflect a lack of compassion and understanding. Patients should be entitled to ask for a replacement who is less anxious or prone to panic, but they cannot force other persons to undergo extreme stress against their wishes.[6]

Recognizing that health-care workers have other roles, and as such, may have deep moral beliefs that conflict with what the patient is requesting, as well as understanding that violating these deeply held moral beliefs would be a terrible imposition on the moral integrity of the physician, patients ought to have an obligation in many situations to tolerate a physician's refusal to provide the requested service. This would avoid the imposition of extreme stress on the physician that in many cases is equal to, if not greater than, that associated with physical risk. Further, as Sokol noted, this obligation is strengthened in light of the fact that patients will generally be able to locate an alternative provider who will not face the same moral "fears."

For these reasons, patients should be encouraged to check the registry of unwilling physicians before seeking care and to select only those doctors who match their own preferences. However, not all patients will immediately be aware of this registry, and even if they are, it is likely too harsh to make them bear the consequences of failure to match by allowing refusing physicians to withhold relevant information, for example. Further, seeking out a morally compatible provider may not be an option in an emergency and may be too much to expect of the young or vulnerable patient, not to mention that even a patient who was previously prepared to accept limitations on the services that her physician is willing to provide may change her mind down the road. Thus, whenever feasible, patients should try to avoid or resolve conflict situations on their own, but their ability to do so should not change or eliminate the refusing physician's baseline duties, with the possible exception of referral to a willing provider.

Notice of Moral Beliefs

Notice of a physician's moral beliefs and refusals is integral to doctor-patient matching and avoidance of many instances of conflicts of conscience, and it is in the patient's best interest to make sure that this information comes to light at the earliest possible juncture. Revelation of a potential mismatch will allow the patient to seek an alternative provider before investing a substantial amount of time and energy in developing a relationship with the refuser, and before a time-sensitive situation arises that could inhibit the search for an alternative physician.[7] Further, should another provider not be available when the patient actually seeks the service in question, she may be able to obtain only financial compensation and not the service itself. Thus, even if the physician fails to broach the topic first, patients have compelling reasons to check the registry of conscientious refusers whenever they can before embarking on a new doctor-patient relationship and to discuss these matters with potential and current physicians. As David Orentlicher notes, patients should take it upon themselves to discover their physician's values,[8] and in light of increasing media attention to how the religious beliefs of doctors influence their practice, some patients are doing just that.[9]

Nevertheless, many patients may not consider the possibility that any given physician would refuse to perform a particular service falling within his or her specialty. Especially before knowledge of the licensing board solution has become widespread, it is rather unlikely that a patient will have researched the physician's refusal status prior to seeking care. Thus, in light of the importance of this information to patients, and because patients may not yet be "accustomed to questioning their physicians about their values and may be uncomfortable with the idea, it is important that physicians volunteer the information and not wait for their patients to request it."[10] It is the refusing physician's obligation to ensure that the potential patient is aware of the nature of his or her refusals; notification of the licensing board alone is not enough.[11] Additionally, if a physician who was not initially opposed to a given service changes position, that physician must notify all existing patients of this change so that they can promptly decide whether they would prefer to find a new provider.

Some commentators put things more cynically than this, arguing, for example, that "practitioners who would place their own interests before care for their patients' interests should at least provide potential patients with prior warning of their intention to resolve this conflict in their own favour."[12] But disdain for the refuser's choice aside, there is general agreement that whenever possible, prior notice is an absolute prerequisite to any legitimate conscientious refusal; even refusers themselves have no reason to balk at this minimal obligation. Importantly, the value of this sort of notice requirement has been discussed in the pharmacy context,[13] and some courts have ordered hospitals to comply with a patient's end-of-life wishes despite their incompatibility with the institution's conscience precisely because the institution failed to make its policies known from the outset of a patient's care.[14] Disclosure of the physician's refusal status should be considered an integral part of informed consent—a species of preconsent to the scope of the doctor-patient relationship. However, once that relationship is established, other sorts of more specific disclosures will be necessary as well.

Informed Consent

While informed consent in medicine is a relatively new development associated with the shift from physician paternalism to patient autonomy, the obligation to disclose relevant information to patients has been recognized by the courts, professional societies, and government agencies for the past several decades. In 1972, the seminal case of *Canterbury v. Spence* established that physicians must communicate the information that patients need to know in order to make informed decisions about their medical care, measured by information that would be likely to influence the decision of a reasonable person.[15] Just a few years later, President Reagan's Commission for the Study of Ethical Problems in Medicine and Biomedical and Behavioral Research issued a report on informed consent, stating that doctors have an obligation to disclose all reasonable treatment options to their patients, "including those [the doctor] does not provide or favor, so long as they are supported by respectable medical opinion."[16] Similarly, the AMA has stated that the "patient's right of self-decision can be effectively exercised only if the patient possesses enough information to make an intelligent choice.... The physician has an ethical obligation to help the patient make choices from among the therapeutic alternatives consistent with good medical practice."[17] Most commentators agree and have appropriately rejected the idea that physicians' personal morality should be permitted to hinder their professional duty to obtain full informed consent from their patients.[18]

Importantly, this obligation is entirely separate from any obligation to personally provide the services disclosed. For example, in a case introduced in chapter 1, a physician and nurse-midwife refused to terminate the pregnancy of a woman at twenty weeks gestation suffering from a premature rupture of her membranes. The patient was given the option of remaining under the care of the refusing providers to deliver the fetus or being transferred to another facility and another physician to receive an abortion. The patient chose the former, delivered a dead fetus, suffered hemorrhaging and septic shock, and ultimately lost her uterus. She won her malpractice case against the physician and nurse-midwife, despite the fact that Pennsylvania has conscience clause protection for "all persons who refuse to obtain, receive, subsidize, accept or provide abortions including those persons who are engaged in the delivery of medical services and medical care."[19] The malpractice liability, however, did not stem from the refusal to provide the abortion, but instead was a result of the failure of informed consent—the patient was not told that by waiting to deliver the fetus or to obtain an abortion she was at increased risk for life-threatening infection and was in fact already showing the signs of one.[20]

This court got things right, but it may have reached a different conclusion had the state's conscience clause been broader, as it is in Mississippi, where the relevant statute provides that a "health care provider has the right not to participate, and no health care provider shall be required to participate in a health care service that violates his or her conscience."[21] The term "participate" is specifically defined to include *counseling*, *advising*, performing, assisting in, referring for, admitting for the purposes of providing, or participating in providing any health-care service.[22] Under Mississippi's expansive language, a physician might without consequence withhold even relevant and medically accurate information that the patient needs to make an informed decision about his or her care and treatment.

While Mississippi's statute seems open to an interpretation that could be used to override traditional informed consent principles, that goes too far, for those principles are an essential aspect of the social negotiation between doctors and the public. Patients are often uniquely vulnerable to their doctors for information about their medical options,[23] and if their doctors do not provide that information, patients may not even realize that they are missing out, seriously hindering their ability to protect themselves. After all, how can you fill a gap that you are unaware of? Allowing such behavior by refusers would be fundamentally at odds with the trust lay persons are forced to place in their physicians as a result of information asymmetry. Thus, permitting this violation of duty on grounds of conscience could transform the balanced solution aimed at protecting both doctors and their patients into one that offers too much protection to doctors at too great an expense to patients. Informed consent must be treated as an integral obligation of each and every person wishing to become and remain a physician.[24]

Just like nonrefusers, refusing physicians should be expected to disclose all of the treatment options and information about those options that the reasonable patient would find material to his or her decision-making process.[25] However, the physician's duty of disclosure is not unlimited and is usually confined to the provision of information about reasonable alternatives that are currently part of the standard of care. Accordingly, if the intervention to which the physician is morally opposed is not medically indicated for a particular patient under his present circumstances or is so technologically new that it is not yet a medically accepted practice, the physician might legally refrain from disclosure. Because this is such a professionally driven standard, however, perpetuation of any existing, but inappropriate, failures of information within the medical profession, or "conspiracies of silence," must be avoided.[26] Thus, materiality to the patient should be interpreted quite broadly both in the context of dealing with conscientious refusals and more generally.

Further, despite the increasing ability of patients to access medical information on their own, physicians ought to be expected to offer material information unless it is apparent that *this* patient is already aware of the relevant options, lest physicians overestimate the sophistication or knowledge of those under their care.[27] Also, it should not be assumed that all patients understand what medical services a Catholic physician will refuse to provide, for example, or that a patient who originally accepted the limitations on services provided by her physician continues to accept those limitations when faced with a personal change of circumstances. Therefore, to satisfy their informed consent and notice obligations, physicians must disclose their specific moral limitations at the initiation of the doctor-patient relationship and reiterate them if and when those limitations become relevant as options for the patient.

Notably, this is not the sort of obligation that patients could mitigate by seeking a different physician, since physicians will owe this informational responsibility to each and every patient under their care. For example, even if a physician told a patient from the start that he will not perform sterilization services, and even if the patient chose to remain under the physician's care, the physician must disclose sterilization as an option should the patient later suffer from a condition that would make a future pregnancy potentially fatal. The right of informed consent is so fundamental to patient autonomy that it should be considered waived only based on a patient's clearest statements to that effect at the time the information at issue would be relevant.[28]

Additionally, the physician must make it clear that the refusal to provide the disclosed service is based on moral, not medical reasons, since failure to make that distinction could result in patients mistakenly concluding that the physician has physiological, evidence-based, objective grounds for refusal. Patients might then fail to seek the desired service elsewhere even though it may have been their preferred option had they been fully informed. Along similar lines, objecting physicians must be certain to provide unbiased clinical information about the patient's treatment options without selectively excluding information or framing it in a particular way, in terms of complication probabilities rather than success rates, for example, in order to avoid improperly influencing the patient to comply with the physician's wishes.[29] At the very least, personal bias should be identified as such, so that the patient is not misled into believing that the physician's advice reflects his or her expert judgment as to how to best accomplish the patient's objectives.[30] We will see subsequently, however, that conscientious refusers need not be limited solely to clinical disclosures.

While certainly less disagreeable than demanding that the refusing physician actually provide the objectionable service, refusers might nevertheless argue that simply informing patients of the availability of these services makes them impermissibly complicit in wrongdoing, despite the fact that someone else makes the final decision to engage in such wrong-doing or performs the act itself. Although claims of moral complicity are not to be taken lightly, this line of reasoning cannot overcome the serious harms to patients that can result from being kept in the dark. More importantly, even if the service in question is not integral to a patient's well-being such that nondisclosure would not likely result in any signifi-cant physical harm, doctors simply lack any authority to withhold or slant information for fear that patients might make a decision the doctor finds personally unacceptable.

This potentially deceitful paternalism cannot be tolerated; note that this was precisely the problem that led to the widespread rejection and prohibition of HMO confidentiality agreements—"gag" clauses—that attempted to prevent physicians from discussing treatment options that were not covered by a patient's insurance plan or that were strongly dis-couraged by the plan due to their expense.[31] As autonomous persons, patients must be allowed to make medical decisions for themselves, and the disclosure of relevant information is absolutely essential to this end. Even more than patient autonomy may be at stake in the emergency con-text, however.

Emergencies

Although a physician's obligations are normally contingent on her spe-cialty, competence, and competing obligations, even within the otherwise contentious conscience clause debate there is a significant level of agree-ment that a doctor's religious or moral convictions can never justify endangering a patient's physical safety.[32] The AMA's Code of Medical Ethics explains that the "greater medical necessity of a service engen-ders a stronger obligation to treat,"[33] and some existing conscience clause statutes even void refusal rights in cases of emergency.[34] Further, we saw in chapter 1 that the Emergency Medical Treatment and Active Labor Act (EMTALA)[35] requires nearly every hospital with an emer-gency room to stabilize patients presenting with an emergency medical condition without regard to their ability to pay. The statute preempts inconsistent provisions of state law, including conscience clauses,[36] but because it focuses on institutional obligations of the hospital, rather than individual responsibilities of emergency room physicians, it likely fails to preempt state conscience clauses in their protection of individual doctors. Nevertheless, the statute gives legal recognition to the notion that emergency medical situations can render otherwise acceptable behavior, such as offering a service only to those who will pay for it, unacceptable.

In some—perhaps even most—emergency situations, the physician expressing a normally permissible moral refusal to the service the patient needs will be able to pass the patient along to a willing provider without imposing any additional risk on the patient, such as to another doctor in the same emergency room. That is entirely unproblematic since the patient is made no worse off, and in fact, getting the patient to some willing provider so that the conscientious refuser need not betray his or her moral integrity has been a central goal of this project. So despite the fact that the patient may be in serious medical danger as a general matter, she is not in danger from this sort of refusal and these are not the sorts of emergencies that concern us here. Instead, we are concerned with other circumstances where transfer is not possible—extreme variants of the hard cases discussed in the preceding chapter in which the patient is entirely dependent on *this* physician for care, even more so than in other cases where the patient will at least have the time to seek care elsewhere, though perhaps at some faraway location. In these more extreme situations, there is no time, no recourse, no real alternative.

A physician's refusal in this sort of emergency does precisely what the proposal explored throughout this book tries to avoid—it undeniably imposes the physician's personal beliefs on the patient. The patient denied non-emergency services from the last doctor in town has also been seriously wronged, but her opportunity is not necessarily inevitably and forever lost. Although she may have to travel some distance and incur financial consequences that make access to the desired service difficult, and these burdens are wrongfully imposed, such obstacles can at least theoretically be overcome. That is simply not the case here, where time is the single decisive factor, and the consequences of denied access may be permanent and grave. These vital considerations, combined with the comfort offered by the doctrine of double effect (and even potential misunderstandings of it), may explain the widespread willingness to nullify in these emergency situations whatever acceptance of conscientious refusal may exist in other contexts.

The doctrine of double effect, initially proposed by Thomas Aquinas as a justification for the permissibility of killing another through an act of self-defense, refers to the notion that one is responsible only for the intended consequences of one's actions, even if other consequences occur and were in fact foreseeable. It provides special permission for *incidentally* causing a morally bad result as a side effect of a good and permissible end that is sufficiently good to outweigh the harms associated with the side effect. The doctrine has been particularly useful in the context of palliative care, where physicians are widely considered to be morally without fault for alleviating a patient's pain by (with the patient's consent) administering drugs such as morphine in doses that will likely shorten life, so long as they intended to relieve pain and only foresaw, but did not intend, the hastening of death.[37]

Importantly, however, the doctrine absolutely does not give permission to undertake the ethically problematic act as a *means* to a good end, and that is where much of the confusion lies. Thus, the palliative care example works only because hastening death is not itself the method of relieving the patient's pain—the doctors are not trying to put the patient out of his misery by killing him, and they could not permissibly, under this doctrine anyway, inject drugs with the direct intention of terminating the patient's life. Instead, the painkillers are the morally neutral means of relieving the patient's pain, though they have the unfortunate and unintended side effect of earlier death. Similarly, a physician could permissibly remove the cancerous uterus of a pregnant woman in order to save her life, even if it would incidentally result in the death of the fetus. This is because the fetus is not being aborted as a means of saving the mother—extraction of the uterus is the means of accomplishing that goal. Consequences to the fetus are foreseen, but not intended, rendering them an acceptable though unfortunate side effect.

Given this caveat about means versus mere side effects, however, the doctrine of double effect is actually quite limited in its ability to absolve physician behavior in emergencies; it cannot morally excuse all harms simply because those harms were not the agent's ultimate aims. For example, aborting the fetus *because* pregnancy itself threatens the life of the mother would in fact be the means to saving her, and thus would not be merely incidental to the accomplishment of the intended goal. While this sort of abortion may be justifiable on other grounds, it could not be explained using this doctrine, even though it has the same final outcome as the hysterectomy case.[38]

In an emergency scenario, the physician's intentions may be genuinely different, which may simply result in fewer objectors under these circumstances. However, many physicians (and others) are not convinced by the doctrine of double effect, mainly as a result of their desire to avoid bad consequences whether they were intended or not. Others may not even need it to justify their willingness to provide otherwise objectionable services in the face of an emergency because they believe they are doing more good than bad. In any event, where the patient lacks an alternative, physicians ought to have a duty to provide emergency services whether the emergency circumstances assuage their personal conscience or not. In fact, while the obligations of moral refusers are our exclusive focus here, this obligation should likely apply to and be enforced against all physicians with the capacity to help, not just those who would normally conscientiously refuse.[39]

This responsibility, however, should not be extended to such a degree that any medical service that a patient cannot gain access to within the relevant geographic area defined by the licensing board becomes classified as an emergency. The provision of Botox injections will never be mandated under this banner, nor will any other purely cosmetic procedure, even though licensing boards may bear responsibility for ensuring access to a sufficient number of physicians willing to provide these services. On the other hand, if a woman will die without the immediate termination of her pregnancy and another doctor is not readily available to perform an abortion in the requisite amount of time, that service should clearly fall within the realm of emergency responsibilities being set out here. EMTALA, for example, defines an emergency medical condition as one that manifests itself in the patient by "acute symptoms of sufficient severity ... such that absence of immediate medical attention could reasonably be expected to" place the patient's health in serious jeopardy or result in serious impairment of bodily functions or dysfunction of a bodily organ or part.[40]

Between these extremes, however, will be services that present a closer call in the sense that they are generally not emergencies as we tend to traditionally define that term, but if put off too long, will eventually be foreclosed as opportunities, resulting in serious negative consequences. Nontherapeutic abortion is a good example, as is the cryopreservation of embryos prior to cancer treatment. Similarly, the patient denied the morning-after pill will not die, but if she cannot access it within three days, it is too late; in these cases, access delayed is access denied. A woman seeking an abortion for purely financial reasons may not be in an "emergency" situation at two months, but could be at six, after which point she may have no legal choice but to carry the pregnancy to term. Must the doctor who sees her the day before her third trimester begins provide the abortion or abandon the legal protection of his conscience that would otherwise exist? Or would abortion under such circumstances never rise to the level of a medical emergency, since the woman's physical health is not threatened?

Situations that will eventually develop into indisputable emergencies but that do not presently threaten the patient's life are even more challenging, since it is not clear when the emergency threshold is crossed. Consider the case of Kathleen Hutchins, whose water broke at the start of her second trimester, rendering miscarriage a certainty within weeks. Those weeks, however, created a significant risk of infection that could have led to infertility or death. Her physician explained the problem in detail, unlike the providers in the similar Pennsylvania case discussed above, and ultimately recommended abortion. However, the local hospital would not permit the procedure until the patient actually suffered from the infection.[41] Had her physician refused and no willing doctor was available to step in, when would emergency obligations have been triggered?

These middle-ground cases demonstrate that emergency services are subject to a wide variety of definitions that would impose a wide variety of obligations on physicians, whether they hold conscientious objections or not. At the very least, where a physician's refusal will impose an avoidable, serious, irreversible consequence on a patient at this moment or in the very near future because the patient truly could not get the care she needs elsewhere, it seems reasonable to impose emergency obligations to provide that care.[42] In those circumstances, the consequences to the patient are sufficiently extreme to justify enforcing the physician's duty through the forfeiture of conscience clause protection, even if other considerations won out in the non-emergency hard case.

In fact, some may even suggest that these consequences are so grave as to demand that all doctors gain the competency to prepare themselves for emergency situations arising in their specialty. However, we have already seen that this argument sacrifices the compromise approach, and that goes too far, eliminating too much physician freedom and moral diversity. Moreover, there is currently no guarantee that everyone in need of emergency care will have access to a competent physician in all circumstances. Properly implemented, the institutional solution should result in reasonable access, so the individual facing an emergency situation for a controversial procedure should be no less likely to gain access to competent emergency care than the general population in need of non-controversial emergency services. Thus, it is sufficient to limit emergency obligations to those already competent to provide the service the patient seeks. This may leave some patients without a physician in a true emergency—and even with this obligation and threatened penalty, some competent doctors will nevertheless conscientiously refuse—but that might be the case anyway, and no compromise solution can fix every problem for every party. This approach is a step forward, despite its limitations.

The scenarios at the margins, however, pose tougher questions. Deciding whether the serious foreclosed-opportunity or pre-emergency cases are—or more accurately, at what point they should be treated as—true emergencies that carry emergency obligations and penalties rather than just regular hard cases in which the physician has a duty that will not be legally enforced will be quite difficult. It will require a comparative assessment of the sorts of consequences likely for patients who are denied care in these contexts and those considerations that counseled against enforcing physicians' nonemergency hard-case obligations. There is clearly much room for continuing debate regarding these marginal cases, offering another example of the sort of line drawing that is best left to licensing boards with input from stakeholders standing outside of the profession.

Referral

Many commentators would impose one additional obligation on physicians even if they are willing to permit doctors to refuse to comply with patient requests on moral grounds: a requirement that refusers give the patient "reasonable assistance and sufficient opportunity to make alternative arrangements for care,"[43] and perhaps even refer the patient directly to a willing provider. While referral requirements would clearly be problematic if there is no one to refer to, that problem will be avoided (at least eventually) through the licensing boards' responsibility to ensure an adequate supply of nonrefusers. However, imposing a duty to refer may nevertheless be more difficult to defend than the other baseline responsibilities explored in this chapter.

This difficulty stems from the fact that each of these other physician obligations could be justified on the grounds that it either imposed an absolutely minimal burden on the refusing physician or, if it involved a greater burden such as complicity or actual provision of the service, that burden was essential to protection of the patient's autonomy, safety, or serious medical interests. While awareness of how to access a desired service is also crucial, since forcing licensing boards to ensure an adequate supply of willing physicians will do little good if patients have no idea that they exist, who they are, or how to connect with them, the critical difference is that the patient could likely protect his or her own interests in this context. Unlike the patient who is kept in the dark about possible options or the patient facing emergency circumstances, the patient whose treatment alternatives are open, obvious, and not time-sensitive could reasonably seek information about willing physicians from a source other than the refuser—namely, from the licensing board. Therefore, imposing the serious moral burdens of complicity by demanding that refusers refer the patient to a willing provider may be uncalled for.

If licensing boards must maintain lists of registered refusers, they could easily create lists of presumably willing providers based on the remainder of licensed physicians. Rather than depending on physicians for this information, why not eradicate the need for their complicity in these circumstances by simply expecting patients who have been denied services by one doctor to self-refer to some other willing physician on the list? To facilitate such behavior, licensing boards could distribute this information, organized by geographic area, to all residents of that area. While there might be some concern that this amount of information would be overwhelming, particularly because all potential patients would have to be provided with information about all potential refusers and their alternates, it need be no more overwhelming than a phone book. More practically speaking, however, the provision of these materials would prove quite expensive for licensing boards, especially considering they would likely call for frequent revision. Further, such lists could be easily misplaced or lost so that they might be of little use when the patient is actually seeking another provider, and there is also the possibility that the lists may not be received at all.

The better option would be for licensing boards to take action to alert patients to the existence of a hotline referral service, a website list of willing providers, or some similar source of information. Unfortunately, however, there is no way to be sure that all patients will be aware of these possibilities unless the refusing physicians themselves have an obligation to provide personal notice. In fact, this problem has been noted with regard to laws requiring patients to be permitted to obtain services directly from willing providers by bypassing their primary care "gatekeeper,"[44] as well as Congress's decision to allow Medicaid patients to obtain family planning services outside of their managed care networks if those networks refuse to provide such services.[45] While self-referral appears helpful to resolving problems of complicity in theory, the major difficulty with this solution is that patients frequently may not know they have this power or how to exercise it. Should this concern suffice to require at least some form of referral?

Under current law outside of the refusal context, once a patient engages a physician to treat a condition, that physician must give the patient all necessary and continued attention for as long as that condition requires. However, as we have seen previously, to avoid perpetration of abandonment under legal standards, the physician must simply avoid unilaterally terminating the professional relationship before giving the patient "reasonable notice of withdrawal and affording the patient a reasonable opportunity to seek treatment from another health care provider."[46] Importantly, physicians have no obligation to provide the patient with outright referral to a replacement or to offer any other assistance. Instead, the burden is on the patient to procure another health-care provider if he or she so desires, although the patient's ability to do so will likely affect what counts as reasonable notice on the part of the refusing physician.

Abandonment jurisprudence seems to favor refusing physicians in this regard, and a few outspoken observers similarly reject any and all referral requirements in the face of conflicts of conscience. For example, one conscientious refuser put things quite bluntly, arguing that telling a patient where to obtain a service that the physician refuses to provide is like saying, "I don't kill people myself but let me tell you about the guy down the street who does."[47] In 1993, Edmund Pellegrino seemed to disagree, stating that when possible, a physician who refuses services on grounds of conscience should transfer the patient to another willing colleague, flagging but not resolving the issue of moral complicity this suggestion would entail.[48] Notably, however, he contradicted these sentiments in later commentary, asserting that "respect for the patient's autonomy does not include referral to a physician who will carry out the procedure if that procedure involves an act the physician deems intrinsically and seriously wrong. For a conscientious physician, this would be an inadmissible degree of formal cooperation."[49]

Thus, active assistance in locating an alternative physician willing to provide the service in question is entirely impermissible from Pellegrino's current perspective. Instead, he would impose the burden of arranging transfer on the patient, family, and social services, once the physician has "respectfully, courteously, but definitively" informed the patient of the moral conflict and consequent refusal.[50] He does, however, argue that the physician must at least continue to care for the patient "in accord with the physician's deepest moral beliefs"[51] until a substitute can be found, and at that point must transfer all information, findings, and records to the alternative provider.[52]

However, the no-referral rule propagated by Pellegrino and the common law seems to stand against the weight of much existing moral commentary on this issue. For example, nearly twenty-five years ago, Reagan's Presidential Bioethics Commission concluded that physicians who refuse a patient's request should at least refer the patient to a willing provider.[53] Similarly, in response to a 2006 request by its Medical Student Section asking the AMA to support a policy of prompt and appropriate referrals following conscientious refusal,[54] the AMA's Council on Ethical and Judicial Affairs concluded that "a conscientious objection should, under most circumstances, be accompanied by a referral to another physician or health care facility."[55] The American College of Obstetricians and Gynecologists backs referral even more forcefully, stating that moral refusers have an obligation to "refer patients in a timely manner to other providers if they do not feel that they can in conscience provide the standard reproductive services that their patients request."[56] The secretary of the Department of Health and Human Services, however, maintains that enforcement of this 2007 committee opinion by federally funded entities would violate laws against discrimination.[57]

The American Bar Association has weighed in as well, asserting that patients "cannot make informed decisions unless their health care providers offer complete, accurate, unbiased, and timely information ... about *how* alternative treatments may be accessed".[58] The ABA's Section of Individual Rights and Responsibilities went on to state that conflict of conscience notwithstanding, "no health care professional should be exempt ... from making appropriate referrals."[59]

Other commentators echo one another, arguing that the privilege of conscientious refusal brings with it the acceptance of consequences and advance planning, including an obligation to ensure alternative arrangements for satisfaction of the patient's request.[60] Some have even gone so far as to claim that it is the "standard view"[61] or "common understanding"[62] for all professions that a necessary condition on the exercise of conscientious refusal is the provision of a suitable and competent referral. In fact, at least one court not relying on the abandonment framework concluded—based on prevailing views of medical ethics, guidelines developed by local medical and bar associations, and state law regarding unprofessional conduct—that a physician can refuse on personal moral grounds to follow a patient's direction, but must be willing to transfer the patient to another provider who will follow the patient's requests.[63]

Clearly, the issue of referral is among the more difficult aspects of the conscience clause debate and has been described by Robert Veatch as a problem that is "absolutely intractable."[64] There are convincing arguments that the referral obligations of conscientious refusers should be no greater than those imposed on physicians who terminate patient relationships on other grounds, particularly because patients aided by licensing boards may be sufficiently capable of protecting themselves in this regard without demanding the moral complicity of physicians. However, there is no need to simply fall back on the status quo, for referral obligations could be changed for all physicians across the board. There is also a reasonable and persuasive concern that without some sort of referral responsibility, some patients, particularly those who are young or otherwise vulnerable, will go without access as a result of conscientious refusals because they are unaware of their options, and that is precisely the problem we are trying to avoid. Thus, it is quite possible that legislatures would be justified in going either way on this question, so long as licensing boards are required to make information about willing providers as accessible as possible.

However, should they determine that reliance on licensing boards alone will not sufficiently safeguard patients against conscientious refusals, as they likely should, legislatures still need not demand outright referral; physicians need not name names. Instead, the physician's complicity could be minimized to a negligible degree by demanding only that the refuser provide patients with copies of the relevant list of willing providers,[65] or at least make them aware of the existence of such a list and how to obtain it, perhaps through the provision of contact information for the board. The refusing physicians providing such information would still be a part of the chain linking the patient to the service that the physician morally rejects, so at least some minimal level of moral complicity would inevitably remain. But outside of relying on licensing boards exclusively, this is the least involvement possible for the refusing physician and is an acceptable second-best solution.

Notably, there is some argument that this approach involves no complicity at all, for if every patient is given this information during the informed consent process, during which the objectionable service must be disclosed along with other alternatives, physicians will have no actual knowledge as to which patients, if any, plan to seek out one of the alternate physicians. Better yet, the list or contact information could be given during the initial notification process at the outset of the doctor-patient relationship or it could be posted prominently in the physician's office, alongside a statement of the physician's refusal policy. It might also be distributed by a willing employee following a physician's refusal or even handed to every patient at check-in. However, this may be the equivalent of turning a blind eye, like handing a recovering alcoholic a list of local bars that he can visit or not, but refusing to call a cab to take him to a specific one—or perhaps more analogously, refusing physicians may see this as no more permissible than posting a list of known hit men. Constructive knowledge may be all that is required for moral complicity.

Those who feel strongly about even this sort of attenuated cooperation argue that some acts are so heinous that any association with them is unacceptable, but even they recognize that we cannot totally separate ourselves from evil and that a lesser degree of association with an evil act decreases culpability. One's proximity to immorality matters.[66] In fact, some are willing to allow cooperation with immorality when the actor does not share the principal's evil intent, the actor is sufficiently removed from performing the evil act, the actor's act of assistance is in itself morally good or indifferent, and the actor has a proportionately serious reason for cooperating.[67] When these criteria are satisfied, as they would be by the sort of referral requirement being discussed here, moral distance is created, which is essential to distinguishing licit from illicit cooperation.[68]

Actually performing an abortion is associated with more culpability than providing mere assistance with the procedure, just as the nurse who assists is more culpable than the friend who suggests abortion as the best alternative, who is more culpable than the physician who refuses to perform the abortion but provides an outright referral, who is more culpable than the refuser who simply provides the patient with a mandatory list of potentially willing providers or refers the patient to the licensing board itself for more information. Even so, patients should be encouraged to help mitigate a physician's complicity whenever they can by seeking care only from willing providers right from the start. In fact, it is possible that as time passes, efforts to make people aware of the licensing board's lists will become more and more successful, potentially rendering any physician referral requirement only temporary.

One additional question raised for legislatures choosing to impose this limited referral obligation on refusers, however, is when the obligation would attach—should doctors also have a duty to individuals who are rejected as patients during the matching process as a result of potential moral conflict? From a legal perspective, the doctor-patient relationship has not yet commenced, and since duties run with relationships, until there is a relationship, the physician has no duties. Just as physicians who are no longer accepting new patients have no obligation to refer prospective patients elsewhere, the same may be true for conscientious refusers.

On the other hand, if the overbooked physician had an easy list of available physicians still taking on patients, it might not be too much to ask for her to provide it, particularly if the reason why referral responsibilities have not been imposed is at least partly because they would prove too burdensome. Providing this sort of list would not be morally problematic for overbooked physicians, though it would be for many conscientious refusers. But a person screened out as a patient as a result of moral beliefs could have just as difficult a time finding an alternative provider as a person who was a patient at the time of refusal. If the consequences are the same, should it really matter that one person was in a relationship when the service was refused and the other was refused prior to a relationship forming?

It may be too dangerous to impose obligations to nonpatients (outside of the emergency case, as we saw above), and for that reason alone, legislatures may choose to limit a refuser's referral responsibilities to existing patients only, if such responsibilities are imposed at all. However, alternative arguments are persuasive as well. Once again, we seem to have reached an issue that should be resolved through a conversation with interested stakeholders.

With regard to referrals, there is no perfect solution able to guarantee that all patients are aware of information about how to access willing providers while simultaneously allowing refusing physicians to completely avoid moral complicity. Unlike the other obligations described in this chapter, however, this one cannot be justified on the grounds that imposing a duty would not burden the refuser or is absolutely necessary for protection of the patient. Thus, it is the duty most susceptible to reasonable debate. If it would drive conscientious refusers out of the profession, that would likely be a sufficient reason to reject it. Otherwise, at least requiring physicians to refer patients to the licensing board for more information on how to access the services they desire seems to strike an acceptable—and likely preferable, though perhaps not essential—balance between personal and professional integrity.

Self-Evaluation and Reason-Giving

Finally, while not the sort of obligation amenable to legal enforcement, it is important to recognize that at a basic level, physicians have a responsibility to take their refusals seriously, refusing only when there is truly no way to accommodate both their own understanding of what is right and the patient's request for medical services. This obligation is related to the validity and sincerity prerequisites of a licensing board accepting a physician's registration of refusal in the first place, but is also more internally located within the physician, particularly because there is no foolproof way to evaluate whether a physician is doing all he can to square his beliefs with the patient's interests. Thus, the refusing physician must "recognize what the limits of legitimate self-interest are and when that set of interests should be set aside in the interests of his patient or vice versa. This is where the central virtue of practical wisdom[, the capacity for deliberation, judgment, and discernment in difficult moral situations,] comes in."[69]

When a physician faces a conflict of conscience in her professional capacity, that physician should be expected to engage in "strong evaluation."[70] This deep self-reflection will involve a close assessment of her moral objections to complying with the patient's request in order to determine whether there is a true, irreconcilable conflict that cannot be overcome in any way other than refusal. Refusing to provide a service that the patient seeks is often a morally grave action, so the physician must recognize that not every ground for refusal is equally important and that carefully selecting the proper junctures at which to "take a morally dissenting stand is crucial if one's exercise of conscience is to be valid and respected."[71] One's conscience should not be used as an unconsidered shield, especially given the effect of conscientious refusals on patient autonomy; the possibility of compromise should be carefully evaluated in each instance where a physician might consider withdrawal. Most importantly, objecting physicians must seriously consider the possibility that they have misjudged the situation in such a way that deeper reflection would uncover that their values are not in fact in conflict with those of the patient, or at least that the conflict can be surmounted.[72]

Further, while conscience itself is subjective, physicians have an obligation to ensure that their judgments rest on factually correct information.[73] For example, a health-care provider who objects to the morning-after pill based on the incorrect notion that the pill will abort an embryo that has already implanted in the uterine lining has an obligation to get the facts straight before refusing on these grounds.[74] Of course, science cannot answer the vast majority of moral questions important to the conscience clause debate, such as when a human life begins, so if the physician's objection is based instead on the fact that the pill prevents implantation, it cannot be attacked as actually wrong. However, a person's conscientious beliefs can change upon consideration of the strong and persistent pressures that may run counter to it and serve as serious checks on one's own beliefs.[75] Even Pellegrino has noted that moral maturity "requires knowing which acts destroy moral integrity and which do not."[76] Therefore, while self-examination may not ultimately result in eliminating a physician's objection, the objection that remains should at the very least be internally consistent and as well supported by reasons as possible, even if not public reasons in the Rawlsian sense.

Additionally, while some commentators worry that allowing or expecting a physician to discuss her reasons for refusal with the requesting patient will invite moral coercion and patient discomfort, there is a compelling argument for encouraging physicians to share the reasons that linger after deep self-reflection, though of course in a respectful manner. It may not suffice to simply state that they refuse on "moral grounds" without explaining what those grounds are, since allowing such bare refusal could diminish one of the most important benefits of preserving room for conscientious refusal in medicine, namely, exposing patients to different ideas about the service in question. Even John Stuart Mill, the ultimate proponent of individual autonomy and liberalism, suggested that despite a person's actions being entirely self-regarding (if that is ever truly possible), "considerations to aid his judgment, exhortations to strengthen his will, may be offered to him, even obtruded on him, by others," so long as he remains the final judge.[77]

Because patients could be exposed to these ideas and arguments from other sources, explanation of a physician's reasons for conscientious refusal may not rise to the level of an actual obligation, but at the very least, it should be tolerated. Without a doubt, patients seeking unbiased clinical care should not be subjected to lectures on the physician's personal moral beliefs, and physicians must take care to communicate their grounds for refusal in a way that does not hinder or diminish the patient's ability or willingness to disagree, particularly in light of the power imbalance inherent in the doctor-patient relationship. However, there is an important difference between moralizing condemnations or proselytizing and a clear exposition of the grounds for a physician's refusal, offered on a take-it-or-leave-it basis to the patient, particularly when the patient can seek the service elsewhere.

Thus, providing reasons is not itself an invasion of patient autonomy, and may even enhance the patient's independence by uncovering the full range of choices and relevant risks, both moral and physical.[78] For example, the physician who refuses to help a patient conceive a child using her dead husband's sperm but explains the basic reasons why he or she feels that this service is morally wrong beyond simply saying no behaves permissibly. In fact, failure to do so may actually be a species of moral abandonment.[79]

However, it is essential to recognize that the exchange of views is a two-way street, not a mere monologue, and as part of his or her self-evaluation, the physician must also be open to an interchange with the patient. In the context of religious objections arising in the lawyer-client relationship, Lesnick argues that it would be inappropriate for the lawyer simply to testify to the strength of her moral convictions, and instead suggests that the lawyer engage the client's moral agency by inviting the client to reflect and participate in dialogue. He notes that perhaps the lawyer will change her mind, having her moral doubts dispelled as she listens to the client's story. Most importantly, Lesnick argues that if the lawyer cannot approach moral counseling from this open-minded stance of potentially changing her own views, she has no right to try to convince the client to change his.[80]

Similarly, a physician has an obligation to take seriously the patient's reasons for requesting the service and the nuances of the patient's individual situation, while explaining his own reasons for objecting. This open-ended conversation may benefit both parties, potentially reaching an agreement amenable to doctor and patient, but at least helping them to understand one another's perspective, potentially easing some of the tension surrounding the conscience clause conflict.[81] Of course, in the end, patients must have the ultimate say about the medical interventions most amenable to their own interests within clinically relevant boundaries. But physicians need not limit themselves to the provision of entirely clinical information and should be permitted to raise their personal moral concerns about what the patient is asking them to do.

While licensing boards bear the brunt of the obligation under this institutional model for addressing conflicts of conscience in medicine, physicians retain obligations of their own and must comply with several prerequisites before they will be entitled to conscience clause protection. First, physicians who object to the provision of services based on wholly irrelevant patient characteristics rather than the moral characteristics of the service itself ought to be denied protection for any consequences arising from their refusal to satisfy patient requests. Similarly, due to the importance and value of doctor-patient matching on the basis of shared moral beliefs, as well as the fact that advanced knowledge of a physician's grounds for refusal will help minimize conflicts, physicians must provide as much disclosure as they can as early in the relationship as possible. Protection of a physician's conscience, no matter how central the grounds for refusal are to the physician's integrity, should not be extended when the physician fails to obtain full informed consent or when a patient faces an emergency. Legislatures may also choose to impose some sort of referral responsibility, but that debate is far from settled. Finally, while it is impossible to legally enforce the obligations of deep self-reflection and open-mindedness, these too should be treated as professional duties of the refusing physician.[82]

These obligations may require actual involvement in objectionable services or at least some level of moral complicity on the part of the conscientious refuser. But they offer the most substantial compromise possible that takes seriously both patient claims to access and physician claims of conscience, especially considering that patients should be expected to mitigate physician burdens whenever they can. Some refusals are acceptable and even play an important role in moral debate, but others simply go too far. If physicians find that they cannot stay on the right side of this heavily limited baseline, then the opponents of conscientious refusal are right—they have no business being in the profession.

10
Addressing Skeptics, a Model Statute, and Conclusions

The preceding chapters have demonstrated that not only is it normatively ideal to simultaneously accommodate the interests of both doctors and patients, allowing physicians to engage in conscientious refusal and preserving patient access to medical services, but also that it is possible to craft a feasible mechanism for doing so. This final chapter offers a model statute to serve as a guide to state legislatures interested in combating the current conscience clause dilemma, which will need to be supplemented by specific regulations adapted to the circumstances faced by various state licensing boards. Without a doubt, this proposal challenges vested political interests and will require a great deal of political force to implement, but it is perhaps no more controversial than the exclusively physician-centric approach already in place or the patient-focused alternatives offered by others. Thus, it must not be casually dismissed.

While comprehensive arguments have been presented in favor of an institutional compromise spearheaded by state licensing boards, some lingering doubts about the propriety of this proposal undoubtedly remain. The most obvious objection is that this solution allows physicians too much leeway and requires too much accommodation on the part of others, thus failing to strike an appropriate balance between competing interests. For example, some critics are likely to maintain that the level of individual negotiation that will be required on the part of patients if physicians are permitted to conscientiously refuse to provide various medical services will impose tremendous transaction costs that patients should not be forced to bear. Patients will not know what services they can rely on getting from any physician within a given specialty, and instead will have to shop around and take the time to make detailed service agreements with their physicians. Further, physicians will have to notify patients should their moral positions change, forcing patients to potentially have to start shopping all over again.

The licensing board solution, however, significantly mitigates these concerns. In this context, patients could simply access the list of willing providers and select a match—very little individual negotiation would actually be required, and so this argument against the compromise solution is unpersuasive.

Others express a potentially more legitimate concern that it would be an unacceptable inconvenience or burden for patients to have to establish relationships with new, unknown, alternative providers should matching fail and a patient actually be refused services by his or her original physician, even if given a direct referral.[1] In fact, some have argued that patients experiencing a major medical crisis should not be asked to develop a new doctor-patient relationship,[2] and recall that the *Requena* court gave great weight to the wishes of a patient who had decided to forgo artificial nutrition and hydration to stay put while she did so.[3] However, while it may be too much to ask a dying patient to change physical locations to carry out her wishes, it might not be terribly burdensome to allow the patient to remain in one place, but to simply have a new physician perform the requested medical service. In fact, the court left that option open.

Thus, even during a medical crisis, a patient's interest in maintaining continuity in the doctor-patient relationship is not the ultimate trump. Further, one would imagine that the relationship a patient would maintain with an objecting physician would be sufficiently strained that starting fresh with an available and willing doctor—a match—would be the more appealing alternative. It is unclear why any rational patient would choose to unnecessarily engage a physician who cannot satisfy the patient's desires in good conscience,[4] particularly if the physician would refuse to provide the service in question regardless of the presence of professional or legal consequences. In fact, we saw that the patient has an obligation to avoid putting the refuser in such a conflict position whenever possible.

This line of reasoning is supplemented by evidence that the doctor-patient relationship is no longer as strong as it was in the past, mainly as a result of managed care systems; even the courts have noted that "although many patients might prefer to be loyal to their doctors, it is, unfortunately, a luxury they can no longer afford."[5] Further, hospitalized patients in major medical centers and teaching hospitals commonly deal with a variety of different doctors due to rotations within services, not to mention that patient care is increasingly being divided among specialists. Thus, patients are already used to switching physicians at relatively frequent intervals. They would not be uniquely inconvenienced by having to establish a relationship with a new physician as a result of conscientious refusal, even if they have to make a separate office visit, wait for available appointments, or travel a longer distance to the alternative provider. Ultimately, it seems, the most important consideration is not which doctor provides the service, but rather that quality care is rendered in a relatively timely and accessible fashion.

Of course, more than mere inconvenience could result if the refusing physician is the only one the patient can afford or access through her insurance plan, but importantly, these issues reflect broader systemic problems. What sound justification is there for placing the burden on the affordable refuser rather than expecting some otherwise willing physician to simply provide lower-priced care or insisting that the insurer add willing providers to its coverage? Surely, refusers must not charge patients for services they fail to render, and they may even have to accept some losses. For example, if a doctor fails to notify a patient that he will not provide prescriptions for contraceptives when she makes her initial appointment and that patient, whose insurer permits only a single routine gynecological exam per year, requests birth control at the end of her visit, the refusing physician has an obligation to ensure that his actions have not made the patient any worse off. This may involve forgoing reimbursement for the services that were provided so that the patient can have another covered visit with someone else who will provide the desired care. Concerted efforts should be made to ensure that the alternatives available to refused patients are financially accessible, but asking refusers to bear the burdens associated with high-cost medical care and fragmented insurance schemes is too much.

A separate concern has been raised by critics who point to young girls or vulnerable women and state that even a single refusal, no matter how open-minded, nonjudgmental, or uncoercive, and regardless of the availability of information regarding alternatives, could suffice to cause such patients to abandon their quest for an abortion or other potentially controversial reproductive health services—a problem that could be made even worse if the physician is allowed to provide the reasons for his refusal.[6] Assuming that this is true, several responses are possible. First, if conscientious refusers become nothing more than a subspecialty of professionals, there is no reason to believe that their refusals will have this effect. Would a patient who walks into a cardiologist's office seeking contraceptives be convinced to abandon her goal if the cardiologist informed her that he does not provide such prescriptions? Even if she would, would that indicate that the cardiologist has an obligation to execute the service?

Second, if a single refusal in the context of other readily available physicians is enough to convince a patient to carry her pregnancy to term, for example, it may be that her mind was not fully made up at the time of her request and that her willingness to proceed with the pregnancy is actually her more considered choice. This is not at all to say that women are incapable of making these decisions for themselves or that someone should always try to talk them out of it. Rather, it is only to say that if anyone can be convinced to abandon any goal so easily, particularly a goal that could still be accomplished with relative ease, it may be appropriate to reconsider whether that was their true goal at all. In other words, genuinely convincing—not coercing—people to change their minds is not an inherently bad thing.

Finally, for those who remain unmoved by these arguments, perhaps the rules for refusing services to highly vulnerable patients should just be different. For those patients who are made vulnerable not simply as a result of access concerns but are vulnerable for other reasons, the physician's obligations may be heightened. Additional responsibilities could include actually assisting the patient in making an appointment with an alternative provider rather than offering only passive referral and being even more vigilant in assuring the patient that people have lots of differing opinions on the service in question, that the physician's perspective is just one of many, and that the final decision remains the patient's. When a patient is refused services, there may be some level of discomfort, embarrassment, or even shame, but each of these dignitary harms can be minimized. The real issue is access to desired services, which this compromise ensures to the greatest extent possible.

However, some critics will remain entirely unconvinced that a compromise solution is appropriate at all. Those who believe that doctors should never have to violate their personal beliefs under any circumstance or that doctors are no more than the patient's agent may be unwilling to budge, although the former group is likely to find this proposal far more tolerable than the latter. However, the licensing board approach does not aim to elevate one side over the other, and is rather based on the foundation that both physicians and patients have compelling claims, that both sides can be relatively comfortably accommodated, and that doing so is actually in society's best interests. Commentators seeking a total victory—an unyielding resolution—are engaging in a completely different conversation, one that is insufficiently nuanced and should be abandoned entirely.

On the other hand, those willing to accept that a compromise is in fact the preferable way of addressing conflicts of conscience in health care, as many are, may nevertheless believe that a different approach is superior, but they are at least working under the appropriate framework. This book offers the sort of detailed suggestions that have been missing from the debate so far, and while it has attempted to demonstrate the inferiority (or lack) of other compromise alternatives when it comes to physicians specifically, in some places it has offered a variety of options that could be tailored according to local preferences; precisely how the profession fulfills its collective obligation can certainly vary from state to state. For example, different states might conclude that more or fewer services fall within the obligation of the medical profession to make available, and could expand or limit the licensing board's obligations accordingly. They might also decide it is best to permit more or fewer types of physician objections, draw wider or narrower geographic boundaries, or implement a more or less stringent threshold of nonnegotiable physician obligations applicable regardless of one's conscientious beliefs. Further discussion as to how we ought to accommodate the competing interests involved in these conflict-of-conscience scenarios is both expected and welcome, but at the very least, we have a new starting point.

Finally, some readers may be skeptical of the ability of licensing boards to carry the weight of this newly imposed burden, given that it would be an entirely novel responsibility. However, even though licensing boards have traditionally dealt with far more modest tasks, they are in no sense beyond the challenge, particularly in light of the options and examples explored here and the fact that they would be provided with legislative guidance and additional funding to accomplish their mandate. Any new government agency, or agency given a broad new charge, faces a daunting undertaking—imagine the overwhelming nature of the FDA's tasks at its inception and with each new congressional delegation of responsibility. Yet, despite occasionally vigorous criticism, the FDA largely satisfies its objectives, and there is no reason to believe that adequately equipped licensing boards could not do the same. In fact, government agencies already address issues of physician supply, and also support medical schools and graduate medical education. The learning curve will be steep and the challenges great, but it is essential to recognize that licensing boards are not being set up for failure—the foundation is there.

One important point that has yet to be addressed, however, is where the funding that must clearly accompany the institutional solution should come from. There are a couple of possible sources, but first, it is essential to point out that funding should not come from refusers alone. While it is true that "acts of conscience are usually accompanied by a willingness to pay some price,"[7] a price should only be exacted when some duty has been breached. Assuming that physicians have complied with their baseline responsibilities, they are not violating any individual professional duty when they refuse outside of the hard case, and so they should not be expected to pay a premium for refusal. We do not want to disincentivize conscientious refusers when they have done nothing wrong, and we have already seen the worrisome consequences that could arise from penalizing refusers should they actually breach their hard-case duties. Perhaps more importantly, a funding structure based purely on penalties would likely be incapable of generating sufficient revenue, let alone providing a constant and predictable revenue stream.

As an alternative, in light of the value of moral diversity in medicine, which is shared by the entire community, one acceptable approach would be for the general public, through the tax system, to bear the financial burden of resolving access problems relating to conscientious refusal and compensating patients failed by the licensing board. This recommendation is up for debate, however, since there are also good arguments suggesting that the financial burden should be borne by increasing licensing fees for all physicians, given that it is the profession's obligation to ensure availability in light of its monopoly power. Perhaps both funding sources ought to be tapped in order to reflect each of these considerations.

However, given that raising revenue is always a difficult political task and that neither increased taxes nor higher licensing fees are likely to be greeted favorably, legislatures may question whether a compromise solution to the conscience clause dilemma is truly worth the expense, normative superiority notwithstanding. When making that determination, it is important to recognize that some of the increased costs associated with the institutional approach will be mitigated by several factors. First, much of the burden will be front-loaded, consisting of first-time start-up decisions that, while complicated, will need only periodic, and less costly, revision once complete. Second, physician refusals, while potentially vast and wide-ranging, are in reality likely to cluster around a few key services like abortion on which licensing boards can focus their attention and resources, at least initially. Finally, and most significantly, the infrastructure created to deal with problems associated with conscientious refusals in terms of measurements and line drawing may also prove useful in the resolution of other serious access problems. Thus, policymakers should have more than sufficient reason to move forward.[8]

Financial concerns aside, which are certainly surmountable even if somewhat problematic, the licensing board compromise described here makes vast improvements over existing conscience clauses, which tend to be disturbingly both over- and under-inclusive. The biggest problem by far is that these laws and regulations fail to ensure adequate patient access following a physician's refusal on grounds of conscience. Additionally, patients may be insufficiently protected against discriminatory moral refusals, despite the existence of anti-discrimination laws and codes of professional ethics that prohibit discrimination on the basis of patient characteristics. The majority of conscience clauses do not list any exceptions to the protections provided for refusing physicians, and while the most common exceptions are for care needed in medical emergencies, only a handful of states include such specific provisions.[9]

Very few conscience clauses explicitly require refusing physicians to provide notice of their moral beliefs to patients in advance, while as many as a dozen extend protection to the refusal to provide information itself, without regard to the decades of jurisprudence and ethical commentary recognizing the importance of informed consent to patient autonomy.[10] Further, most fail to require doctors to provide any reason for their refusal, and those that do suggest such a low evidentiary threshold for proving a moral or religious objection that the requirement becomes a farce that does nothing to prevent abuse of conscience protection.[11] Finally, by offering protection to refusers without exception for bona fide occupational qualifications or consideration of the burden that having to accommodate refusers could place on their employers, many conscience clauses risk inappropriately forcing employers to subsidize physician conscience. For each of these reasons, calls to provide health-care workers with blanket immunity for conscientious refusal are extraordinarily misguided.

In contrast to each of these examples demonstrating the many ways existing laws must be reigned in, there is at least one important way that the protection offered by these statutes should be expanded. As we saw in chapter 1, many conscience clauses are limited to refusals to participate in abortion procedures. Some states extend protection to the refusal to provide contraceptive services or sterilization procedures, and perhaps also to those with moral objections to withholding or withdrawing life-sustaining care, while others go even further. However, considering that a physician may have as strong and legitimate a moral reaction against medical services and procedures not specifically covered by these laws, there appears to be no reasoned way to justify these seemingly arbitrary and ad hoc distinctions.

There has been some debate as to whether conscience clause protection should be extended one procedure at a time, as the grounds for conflict become evident and it is possible to understand the scope of potential objections and their impact on patients, or whether it should be extended all at once now that the breadth of potential conflicts is clear, with limitations to prevent physician refusals from having too negative an impact on patients.[12] With the safeguards offered by the institutional solution, the latter option appears clearly to be the most appropriate—there is no valid reason to differentiate between services. All genuine moral refusals should be accepted under the restrictions described throughout this book, but currently, only Illinois, Mississippi, and Washington are on the right track, despite the other flaws in their conscience clauses, by protecting refusals to provide any medical service to which the physician is morally opposed.[13]

By this point, the arguments in favor of the institutional licensing board compromise have been made as persuasively as they could be. It ultimately appears to be both feasible and right, and it simply remains to be translated into a legislative solution. What follows is a draft statute (box 1)—a working proposal built on the entire preceding discussion and open to debate and responsive commentary yet to come.

This model statute leaves many questions unanswered, and it will often be up to licensing boards themselves to determine the best mechanisms for implementing this new policy, potentially drawing on the numerous suggestions presented in the preceding chapters. As more and more states choose to adopt the institutional solution to the problems associated with conscientious refusal, they will be able to draw on one another's successes and failures, rendering future facilitation of conscience and access easier. The important things for every state to focus on, however, include the need to define the scope of services falling under the board's obligation; to assess supply, demand, and relevant geographic area; to impose an obligation to calibrate supply to demand when these factors are unacceptably mismatched as a result of conscientious refusal; to create a penalty or compensation scheme when this calibration fails and access is denied; and to protect refusing physicians from liability and discrimination so long as they satisfy baseline obligations to patients.

To use a phrase from the world of political science, a "problem window" has been opened and the time is right for action in the conscience clause arena.[14] Sparked by media coverage of a handful of pharmacists' refusals, which in turn sparked media coverage of the broader issue, the public has become interested, and of course, the interest of politicians

Box 1
The Model Protection of Conscience and Patient Access Act

Section 1.   Findings.[1]

A. Every individual has an interest in exercising his or her personal religious and moral beliefs, regardless of his or her profession. Further,
maintaining a role for moral diversity and moral integrity within the
professions is beneficial to the public.

B. Members of the public rely on physicians to make a variety of legally
permissible services available to them. Thus, the personal religious and
moral beliefs of physicians should not result in denial of access to these
services.

C. Therefore, the legislature recognizes an obligation on the part of the
state to facilitate both the exercise of personal conscience of physicians
and the public availability of their professional services. Given its control over the profession through its role in the licensure and discipline
of physicians practicing under the [State] Medical Practice Act, the
[licensing board] is the most appropriate institution to bear this obligation on behalf of the state.

Section 2.   Obligations of State Licensing Board.

A. *Generally.* The licensing board shall bear the ultimate responsibility for
ensuring that patient access to medical services is not unreasonably
impaired by the conscientious refusal of physicians. This statute does
not impose any responsibility on the licensing board to eliminate obstacles to patient access stemming from any other source.*

B. *Medical Services.*

1. The medical services for which the licensing board bears the responsibility to ensure that patient access is not unreasonably impaired by
the conscientious refusal of physicians shall include all those medical services that are legally permitted and scientifically proven effective for the patient's desired purpose, provided that the board, after
notice-and-comment rulemaking, may exclude any service for which
there is no generalized demand or that would shock the collective
conscience of the community. Notwithstanding any other provision
of this paragraph, the board shall not propose any rule that would

---

* The legislature will also need to address funding issues in order for the
licensing boards to accomplish their new tasks, but this will normally be
dealt with in appropriations legislation. However, if the legislature determines that the best way to fund this compromise involves increasing licensing fees, a provision to that effect should be included here.

Box 1
(continued)

exclude from its responsibility a medical service that has attained the status of standard of care.

2. The licensing board shall reevaluate the list of medical services for which it bears responsibility at reasonable intervals.

C. *Patient Demand.* The licensing board shall measure patient demand for the medical services for which it bears responsibility under (2)(B) at reasonable intervals.

D. *Physician Supply.*

1. The licensing board shall measure the supply of physicians willing and competent to provide the medical services for which the board bears responsibility under (2)(B). The board shall obtain this information by requiring physicians to register any conscience-based grounds for refusal upon their initial licensure in [State] and upon any change in their conscientious beliefs.

2. Before accepting a physician's registration of refusal, the licensing board shall reasonably verify that the physician's grounds for refusal are sincerely based in his or her genuine moral or religious beliefs. Regardless of the sincerity of the physician's grounds for refusal, the board shall reject registrations of refusal based on invidious discrimination against any patient or group.

3. Based on this registry, the licensing board shall compile reasonably current lists of willing physicians and lists of refusing physicians for each of the medical services for which the board bears responsibility under (2)(B), grouped by the geographic areas established under (2)(E). The board shall post these lists on publicly accessible websites and shall make hard copies available to any patient or physician requesting them. The board shall also undertake reasonable measures to publicize the availability of these lists.

4. The inclusion of a physician on a list of willing providers shall not in itself obligate that physician to accept the transfer of or to provide services to any particular patient.

5. The state does not endorse or assume responsibility for any representation, claim, or act of any physician included on a list of willing providers.[2]

E. *Geographic Area.*

1. The licensing board shall determine, through the process of notice-and-comment rulemaking, the appropriate geographic areas within the state in which an adequate supply of willing and competent physicians must be maintained such that patient access to the medical services for which the board bears responsibility under (2)(B) is not unreasonably impaired by the conscientious refusal of physicians.

Box 1
(continued)

2. The appropriate geographic area may differ for different services, based on consideration of the absolute number of physicians willing and competent to perform a particular medical service, the strength of the patient's need for a given service as measured by the service's impact on the patient's health, the existence of transportation barriers, and any other factors the licensing board deems relevant.
3. The licensing board shall reevaluate the established geographic areas at reasonable intervals and shall take appropriate steps to ensure that the geographic areas are as narrow as possible.

F. *Calibration of Supply and Demand.*
1. Upon discovery that there is an inadequate supply of physicians willing and competent to provide the medical services for which the board bears responsibility under (2)(B) in the relevant geographic area established under (2)(E) such that patient access is being unreasonably impaired as a result of conscientious refusal, the licensing board shall take appropriate steps to recruit willing and competent physicians to the area, or to otherwise ensure the availability of those medical services.
2. As necessary, the licensing board shall work with professional associations, medical schools, and other organizations to increase the number of willing and competent physicians available for recruitment.

G. *Licensing Board Accountability.*
1. The licensing board shall compensate, according to an established schedule, a patient unable, as a result of conscientious refusal, to locate a willing and competent physician within the relevant geographic area established under (2)(E) to provide a medical service for which the licensing board bears responsibility under (2)(B).*
2. Failure of the licensing board to ensure an adequate supply of willing and competent physicians has no bearing on the permissibility of a physician's registered refusal.

H. *Authority of State Licensing Board.* The licensing board is authorized to promulgate regulations necessary to implement the provisions of this statute.

---

* Legislatures should also prescribe statutory compensation amounts here, or may delegate this task to some other appropriate body. Further, legislatures should delay the effective date of the licensing board's accountability to patients for some period following enactment of the statute so as to allow licensing boards a reasonable opportunity to satisfy their mandate.

Box 1
(continued)

Section 3.   Protection of Physician Conscience.

A. *Immunity from Liability.* No physician shall be civilly or criminally liable to any person, estate, public or private entity, or public official by reason of his or her refusal to perform any medical service that is contrary to the conscience of that physician.[3]

B. *Discrimination.*
  1. It shall be unlawful for any licensing board or publicly funded medical education program to withhold or revoke the license of, deny admission to, expel, or otherwise discipline any physician or prospective physician based on his or her refusal to perform a medical service that is contrary to his or her conscience, unless the licensing board has determined, through the process of notice-and-comment rulemaking, that the medical service is an essential skill that every professional in a particular specialty must be willing to provide. The definition of essential skills shall give consideration to the goal of permitting sub-specialization by conscientious refusers.
  2. It shall be unlawful for any employer, employment agency, or labor organization to discriminate against a physician based on his or her refusal to perform a medical service that is contrary to his or her conscience in regard to job application procedures, hiring, advancement, discharge, compensation, job training, or any other terms, conditions, and privileges of employment, unless:
     a. The physician's willingness to perform the service is a bona fide occupational qualification reasonably necessary to the normal operation of that employer's, agency's, or organization's particular business or enterprise; or
     b. The employer, agency, or organization demonstrates that it is unable to reasonably accommodate the physician's conscience without undue hardship on the conduct of the employer's, agency's, or organization's particular business or enterprise. "Undue hardship" means an action requiring significant difficulty or expense, in light of the nature and cost of the accommodation, the overall financial resources of the employer, agency, or organization, the number of persons employed, and the effect on expenses and resources of such accommodation upon the employer, agency, or organization.[4]

C. *Exceptions.* Notwithstanding any other provision of this section, a physician shall not be immune from liability or discrimination under any of the following circumstances:

Box 1
(continued)

1. The physician engages in conscientious refusal on the basis of a patient's race, color, national origin, ethnicity, gender, religion, creed, sexual orientation, marital status, or other characteristic that the licensing board determines to be invidiously discriminatory;
2. The physician fails to register with the licensing board his or her conscientious refusal to perform a medical service for which the board bears responsibility under (2)(B);
3. The physician fails to specifically and effectively notify patients, employers, employment agencies, or labor organizations of his or her conscientious refusal to perform particular medical services, despite the feasibility of such notice;
4. The physician fails to disclose material information regarding treatment options to patients because the physician finds those options to be morally or religiously objectionable;
5. The physician engages in conscientious refusal despite the fact that the patient is facing an emergency, as defined by the licensing board through notice-and-comment rulemaking;
6. The physician engaging in conscientious refusal fails to:
   a. Directly refer patients to a willing physician in the relevant geographic area established under (2)(E);
   b. Provide patients with a list of willing physicians in the relevant geographic area established under (2)(E), derived from the licensing board's list of registered refusers; or
   c. Direct patients to the licensing board itself for such information.*

Notes

1. See, for example, Wash. Rev. Code § 70.47.160 (West 2006) ("The legislature recognizes that every individual possesses a fundamental right to exercise their religious beliefs and conscience. The legislature further recognizes that in developing public policy, conflicting religious and moral beliefs must be respected. Therefore, while recognizing the right of conscientious objection to participating in specific health services, the state shall also recognize the right of individuals enrolled with the basic health plan to receive the full range of services covered under the basic health plan").

* This limited referral requirement might be excised by legislatures who determine that patients are sufficiently able to locate an alternative willing physician on their own. Should legislatures choose to impose a referral obligation, whether it applies with regard only to patients or also to prospective patients is a matter for notice-and-comment rulemaking.

Box 1
(continued)

> 2. Sections 2(D)(3), (4), and (5) are based on the Texas Advance Directive Act, § 166.053 (1999).
> 3. Derived from 745 Ill. Comp. Stat. Ann. 70/4 (West 2006).
> 4. Derived from a combination of Title VII and the Americans with Disabilities Act.

never lags far behind. The first step has already been accomplished and these conflicts are clearly on the political agenda, as demonstrated by the number and variety of bills dealing with moral refusals in the health-care setting recently introduced in state legislatures around the nation. Throughout this book, we have developed a balanced solution for the problem as it applies to physicians, but given the incredible salience and divisiveness of moral issues like this one, as well as the need for true compromise that this proposal calls for, it may not *yet* be an idea whose time has come as a political matter.[15] However, given the deep problems with other possible approaches and the superiority of this one, it is likely an idea whose time *will* come.

"A refusal to be an instrument of another's wishes is very different from trying to prevent another from realizing his or her goals,"[16] and this is precisely why a compromise solution to the conscience clause dilemma is possible. It is indisputable that respect for physician conscience may at times be troublesome for patients, inefficient for the health-care system, and when taken too far, detrimental to medical outcomes.[17] Nevertheless, physicians' moral beliefs and values deserve to be taken seriously in both their personal and professional lives, particularly considering that physicians "who refuse, withdraw, or disassociate themselves from certain practices or procedures on grounds of conscience may well be among the more thoughtful and effective members of a healthcare team."[18]

In fact, physicians who are uncomfortable crossing the line between what they see as morally permissible and murky or unacceptable actions should often be praised for raising deep ethical issues rather than blindly marching forward and following orders.[19] Those who are willing to do things that they believe to be wrong, even if those things are in reality perfectly morally acceptable, show that morality does not weigh on them as it should, indicating that they may be amenable to actions that are genuinely wrong. For this reason, it is acceptable to try to persuade people to change their conscientious beliefs, but they ought to be respected for continuing to refuse to act to the contrary when they remain unconvinced.[20]

While the seemingly obvious solution to this problem would be to accept as physicians only those people who believe that what they are being asked to do is morally permissible, that raises problems of its own, since their subjective beliefs may themselves be incorrect. Particularly in areas where we cannot separate right from wrong with certainty, the preservation of serious moral debate within and beyond the profession is essential. We must contend with the simple fact that refusers may be right—in many areas, we just cannot know. Additionally, preserving freedom in the choice of one's profession and, with reasonable limitations, in how to conduct oneself as a professional are both important goals in themselves, not to mention that excluding refusers may not even resolve access problems—though it should be clear by now that even if it would, another approach would still be preferable. Thus, protecting conscience appears as a special good for doctors, patients, and society that ought to be given exceptional consideration whenever possible, particularly when conscientious refusal will not result in harm to patient health and well-being. But avoiding that harm is key. Maximal liberty for *all* parties should be the aim, and while current approaches have tended to be all-or-nothing, that need not be the case.

Rather than sacrificing physician conscience, the availability concerns that have played a crucial role in this debate can often be overcome through serious "dialogue, prudent planning, and the exercise of tolerance, imagination, and political will."[21] The best response to the conflict-of-conscience problem is to eliminate such conflicts between doctors and their patients to the extent possible by facilitating matching based on deep moral values. This should be the goal of the medical profession, particularly since it will help reduce any inconvenience and inefficiency that results from allowing physician refusal. However, to make this a reality, we ought to impose an institutional obligation on state licensing boards, standing in as a representation of the medical profession as a whole, to ensure that patients have reasonable access to willing and competent providers and that the profession's collective obligations are borne by particular refusers as infrequently as possible. Holding licensing boards accountable allows individual liberties—for both doctors and patients—to thrive.

Freedom of conscience is certainly among the most revered of American values, but it is not the only value of importance. As others have noted, the "moral necessity of allowing one to follow the dictates of conscience and the social good of making legally sanctioned medical services readily available must be *jointly* preserved."[22] That goal is achievable, and hopefully both patient- and physician-focused critics will see the promise of this unique institutional solution to accomplish it.

Appendix: Statutes, Regulations, and Case Law

Abortions, Refusal to Perform or Participate, Exemptions, Okla. Stat. Ann. Tit. 63 § 1-741(B).

Access to Legal Pharmaceuticals Act, S. 809, 109th Cong.; H.R. 1652, 109th Cong. (proposed, but not passed).

Administrative Procedure Act, 5 U.S.C. § 500 et seq.

Americans with Disabilities Act, 42 U.S.C. §§ 12111–12117.

Basic Health Plan, Health Care Access Act, Right of Providers, Carriers, and Facilities to Refuse to Participate In or Pay for Services for Reason of Conscience or Religion, Wash. Rev. Code § 70.47.160.

Certificate of Need, Criteria Used in Evaluation of Applications, Fla. Admin. Code Ann. r. 59C-1-030.

Certificate of Need, Montana Code Ann. § 50-5-304.

Certificate of Need, Review Criteria, Fla. Stat. Ann. § 408.035.

Church Amendment, 42 U.S.C. § 300a-7.

Crimes and Offenses, Abortion, Legislative Intent, Right of Conscience, 18 Pa. Cons. Stat. § 3202(d).

Criteria for Certificate of Need Review, 28 Pa. Code § 401.4.

Designation of Health Professional(s) Shortage Areas, 42 C.F.R. § 5.2—Appendices A and C, Part I—Geographic Areas.

Determination of Public Need, N.Y. Comp. Codes R. & Regs. tit. 10, § 709.1.

Emergency Medical Treatment and Active Labor Act, 42 U.S.C. § 1395dd.

General Certificate of Need Review Criteria, Ohio Admin. Code § 3701:12–20.

General Requirements, All Pharmacies, Personnel, 5-19-1 R.I. Code R. § 13.1.3.

Health Care Right of Conscience Act, 745 Ill. Comp. Stat. Ann. 70/1-70/14.

Health Care Rights of Conscience, Miss. Code Ann. § 41-107 et seq.

Maternal Health, Abortion, Cal. Health & Safety Code § 123420.

Medical Facility Not Required to Allow Abortions, Nev. Rev. Stat. Ann. § 449.191.

Military Selective Service Act, Deferments and Exemptions from Training and Service, 50 U.S.C. § 456.

Oregon Death with Dignity Act, Or. Rev. Stat. §§ 127.800-127.897.

Performance of Abortions, Voluntary and Informed Consent Required, Information Provided, S.D.C.L. § 34-23A-10.1 (under judicial review at time of press).

Pharmacies' Responsibilities, Wash. Admin. Code § 246-869-010.

Pharmacist's Professional Responsibilities, Wash. Admin. Code § 246-863-095.

Pharmacy Facilities, Ariz. Admin. Code § 4-23-611.

Pharmacy Practice Act, Ill. Admin. Code tit. 68, § 1330.91.

Texas Advance Directives Act, §§ 166.046, 166.052, & 166.053.

Title VII of the Civil Rights Act of 1964, 42 U.S.C. § 2000(e) et seq.

Unlawful to Require Participation in Abortion, Nev. Rev. Stat. Ann. § 632.475.

Unprofessional Conduct, Cal. Bus. & Prof. Code § 733(b)(3).

Case Law

Beal v. Doe, 432 U.S. 438 (1977).

Blue Cross & Blue Shield of Wisconsin v. Marshfield Clinic, 65 F.3d 1406 (7th Cir. 1995).

Brophy v. New England Sinai Hospital, Inc., 497 N.E.2d 626 (Mass. 1986).

Canterbury v. Spence, 464 F.2d 772 (D.C. Cir. 1972).

Catholic Charities of the Diocese of Albany v. Serio, 7 N.Y.3d 510 (2006).

City of Boerne v. Flores, 521 U.S. 507 (1998).

Cobbs v. Grant, 8 Cal.3d 229 (1972).

Cogan v. Harford Memorial Hospital, 843 F. Supp. 1013 (D.Md. 1994).

Conservatorship of Morrison v. Abramovice, 253 Cal. Reptr. 530 (Cal. Ct. App. 1988).

Cruzan v. Director, Missouri Department of Health, 497 U.S. 261 (1990).

Cutter v. Wilkinson, 544 U.S. 709 (2005).

Davis v. U.S., 629 F. Supp. 1 (E.D. Ark. 1986).

Employment Division v. Smith, 494 U.S. 872 (1990).

Endres v. Indiana State Police, 349 F.3d 922 (7th Cir. 2003).

Free v. Holy Cross Hospital, 505 N.E.2d 1188 (Ill. App. Ct. 1987).

FTC v. Freeman Hospital, 69 F.3d 260 (8th Cir. 1995).

FTC v. Tenet Health Care Corp., 186 F.3d 1045 (8th Cir. 1999).

Gillette v. U.S., 401 U.S. 437 (1971).

Gordon v. Lewistown Hospital, 272 F.Supp.2d 393 (M.D. Pa. 2003).

Grace Plaza of Great Neck, Inc. v. Elbaum, 183 A.D.2d 10 (N.Y. App. Div. 1992).

Gray v. Romeo, 697 F. Supp. 580 (D.R.I. 1988).

Griswold v. Connecticut, 381 U.S. 479 (1965).

Harbeson v. Parke-Davis, Inc., 656 P.2d 483 (Wash. 1983) (en banc).

Harris v. McRae, 448 U.S. 297 (1980).

In re the Conservatorship of Helga M. Wanglie, No. PX-91-283 (P. Ct. Minn. Hennepin County June 28, 1991).

In re Evanston Northwestern Healthcare Corp., No. 9315 (FTC Initial Decision Oct. 20, 2005).

In re Jobes, 529 A.2d 434 (N.J. 1987).

In re Requena, 517 A.2d 886 (N.J. Super. Ct. App. Div. 1986).

In re Summers, 325 U.S. 561 (1945).

Jackson v. Metro Edison Co., 419 U.S. 345 (1974).

Johnson v. Robison, 415 U.S. 361 (1973).

Kenny v. Ambulatory Centre of Miami, Fla., Inc., 400 So.2d 1262 (Fla. Dist. Ct. App. 1981).

Kenny v. Wepman, 753 A.2d 924 (R.I. 2000).

Leach v. Drummond Medical Group, 192 Cal. Rptr. 650 (Cal. Ct. App. 1983).

Maher v. Roe, 432 U.S. 464 (1977).

Martin by Scoptur v. Richards, 192 Wis. 2d 156 (1995).

Matthies v. Mastromonaco, 709 A.2d 238 (N.J. Super. Ct. App. Div. 1998), aff'd, 733 A.2d 456 (N.J. 1999).

Moose Lodge No. 107 v. Irvis, 407 U.S. 163 (1972).

North Coast Women's Care Medical Group, Inc. v. Superior Court (a.k.a. *Benitez v. North Coast Women's Care Medical Group*), California Supreme Court Case S142892. For more information, see http://appellatecases.courtinfo.ca.gov and www.lambdalegal.org.

Payton v. Weaver, 182 Cal. Rptr. 225 (Cal. Ct. App. 1982).

Pierce v. Ortho Pharmaceutical Corp., 417 A.2d 505 (N.J. 1980).

Planned Parenthood v. Casey, 505 U.S. 833 (1992).

Planned Parenthood of Minnesota v. Rounds, 467 F.3d 716 (8th Cir. 2006), rehearing en banc, No. 05-3093 (8th Cir. Argued Apr. 11, 2007).

Ricks v. Budge, 64 P.2d 208 (Utah 1937).

Rodriguez v. City of Chicago, 156 F.3d 771 (7th Cir. 1998).

Roe v. Wade, 410 U.S. 113 (1973).

Sard v. Hardy, 281 Md. 432 (1977).

Schloendorff v. Society of New York Hospital, 211 N.Y. 125 (1914).

Shelton v. University of Medicine & Dentistry of New Jersey, 223 F.3d 220 (3rd Cir. 2000).

Sherbert v. Verner, 374 U.S. 398 (1963).

Smith v. Fair Employment & Housing Commission, 913 P.2d 909 (Cal. 1996).

St. Agnes Hospital v. Riddick, 748 F. Supp. 319 (D. Md. 1990).

Stormans v. Selecky, W.D.Wa., Case No. C07-5374RBL, Nov. 8, 2007, www .alliancedefensefund.org/UserDocs/StormansPIRuling.pdf.

Swanner v. Anchorage Equal Rights Commission, 874 P.2d 274 (Alaska 1994).

Swanson v. St. John's Lutheran Hospital, 597 P.2d 702 (Mont. 1979).

Taylor v. St. Vincent's Hospital, 369 F. Supp. 948 (D. Mont. 1973), aff'd, 523 F.2d 75 (1975).

Thomas v. Abdul-Malek, No. 02-1374 (W.D. Pa July 29, 2004).

Trans World Airlines, Inc. v. Hardison, 432 U.S. 63 (1977).

U.S. v. E.I. DuPont de Nemours, 351 U.S. 377 (1956).

U.S. v. Rockford Memorial Corp., 898 F.2d 1278 (7th Cir. 1990).

U.S. v. Seeger, 380 U.S. 164 (1965).

Vandersand v. Wal-Mart Stores, Inc. (being decided by the U.S. District Court for the Central District of Illinois at time of press).

Walker v. Pierce, 560 F.2d 609 (4th Cir. 1977).

Warthen v. Toms River Community Memorial Hospital, 488 A.2d 299 (N.J. Super. Ct. App. Div. 1985).

Welsh v. U.S., 398 U.S. 333 (1970).

Notes

Introduction

1. Erich H. Loewy, "Physicians and Patients: Moral Agency in a Pluralistic World," *J. Med. Human. & Bioethics* 7, no. 1 (1986): 57, 64.

2. Julian Savulescu, "Conscientious Objection in Medicine," *Brit. Med. J.* 332 (2006): 294, 297.

3. Teresa Stanton Collett, "Professional versus Moral Duty: Accepting Appointments in Unjust Civil Cases," *Wake Forest L. Rev.* 32 (1997): 635, 635 (quoting Pope John Paul II, The Gospel of Life, 1995, para. 74).

4. Farr A. Curlin, Ryan E. Lawrence, Marshall H. Chin, and John D. Lantos, "Religion, Conscience, and Controversial Clinical Practices," *New Eng. J. Med.* 356, no. 6 (2007): 593, 597.

5. Sabrina Rubin Erdely, "Doctors' Beliefs Hinder Patient Care," *Self Magazine*, June 2007, www.self.com.

6. Or. Rev. Stat. §§ 127.800–127.897.

7. Or. Rev. Stat. § 127.885.4.01 (2).

8. The AMA Code of Ethics § 2.211 states that physician-assisted suicide is "fundamentally incompatible with the physician's role as healer, would be difficult or impossible to control, and would pose serious societal risks." Thus, it implores physicians to avoid involvement in the practice, while being sure not to abandon patients at the end of their lives.

9. See, for example, Peter Korn, "Doctor's Ethics Run Counter to Hospital Policy," *Portland Tribune*, April 21, 2006. Note that six of the fifteen Oregonian patients who took their own lives in 1998 requested legal medications from more than one physician before finding one who would begin the prescription process. Mark R. Wicclair, "Conscientious Objection in Medicine," *Bioethics* 14, no. 3 (2000): 205, 206. This could have been due to physician concerns regarding liability given worry that the Death with Dignity Act would be judicially overturned, but it could also have been due to personal moral disagreement with

what the patients planned to do or concern about violating established norms of professional ethics that focus on avoiding harm to the patient.

10. Nancy Gibbs, "Defusing the War over the 'Promiscuity' Vaccine," *Time Magazine*, June 21, 2006; Rob Stein, "Cervical Cancer Vaccine Gets Injected with a Social Issue," *Washington Post*, October 31, 2005, A3.

11. National Vaccine Information Center, "Press Release: Merck's Gardasil Vaccine Not Proven Safe for Little Girls," June 27, 2006; "Press Release: Vaccine Safety Group Releases Gardasil Reaction Report," February 21, 2007, www .nvic.org.

12. Bernard M. Dickens, "Preimplantation Genetic Diagnosis and 'Saviour Siblings'," *Int. J. Gynecology & Obstetrics* 88 (2005): 91; "Embryos Screened to Create 'Savior Siblings': Babies Compatible as Stem-Cell Donors," *Milwaukee Journal Sentinel*, May 4, 2004. Note that until 2004, the UK's Human Fertilisation and Embryology Authority policy prohibited the testing of embryos for tissue matching with seriously ill siblings due to concern for the health and well-being of the resulting child, allowing the tests only when there was a risk that the embryo itself would have the same genetic disease as its sibling. Now, HFEA allows such testing, but it is still heavily regulated and will be approved only when the physician treating the seriously ill sibling can attest that this method is being used as a last resort. "Press Release: HFEA Agrees to Extend Policy on Tissue Typing," July 21, 2004. In the United States, the creation of savior siblings is entirely unregulated.

13. David King, "Why We Should Not Permit Embryos to Be Selected as Tissue Donors," *Bull. Med. Ethics* 190 (2003): 13.

14. Sean Murphy, "Service or Servitude: Reflections on Freedom of Conscience for Health Care Workers," 3, www.consciencelaws.org.

15. John K. Davis, "Conscientious Refusal and a Doctor's Right to Quit," *J. Med. & Phil.* 29, no. 1 (2004): 75, 81–82 (original emphasis eliminated).

16. Tom L. Beauchamp and James Childress, *The Principles of Biomedical Ethics*, 5th ed. (Oxford: Oxford University Press, 2001), 38.

17. Beauchamp and Childress, *The Principles of Biomedical Ethics*, 356.

18. For example, David B. Brushwood essentially punts the question, stating broadly that "whatever system is developed for accommodating personal beliefs, it should reflect the legal boundaries of appeals to conscience, and it should ensure that patient needs are being met." "Conscientious Objection and Abortifacient Drugs," *Clinical Therapeutics* 15, no. 1 (1993): 204, 211.

19. Lawrence J. Nelson and Robert M. Nelson, "Ethics and the Provision of Futile, Harmful, or Burdensome Treatment to Children," *Critical Care Med.* 20, no. 3 (1992): 427.

20. Carol S. Weissert and William G. Weissert, *Governing Health: The Politics of Health Policy*, 2nd ed. (Baltimore: Johns Hopkins University Press, 2002), 326–327.

Chapter 1

1. 42 U.S.C. § 300a-7 (1973).

2. *Taylor v. St. Vincent's Hospital*, 369 F. Supp. 948 (D. Mont. 1973), aff'd, 523 F.2d 75 (1975).

3. 42 U.S.C. § 300a-7(b).

4. 42 U.S.C. § 300a-7(c).

5. Though some states permitting the sort of therapeutic abortions needed to protect the health or life of the mother even before *Roe* had previously adopted statutes protecting refusals to provide that service, the Church Amendment was the first of its kind at the federal level. See Martha Swartz, "Conscience Clauses or Unconscionable Clauses: Personal Beliefs versus Professional Responsibilities," *Yale J. Health Pol'y, L., & Ethics* 6 (2006): 269, 281.

6. Claire A. Smearman, "Drawing the Line: The Legal, Ethical, and Public Policy Implications of Refusal Clauses for Pharmacists," *Ariz. L. Rev.* 48 (2006): 469, 477.

7. 745 Ill. Comp. Stat. Ann. 70/3 (West 2006).

8. Miss. Code Ann. § 41-107-3 (2005).

9. Wash. Rev. Code § 70.47.160 (West 2006).

10. For an excellent survey of these statutes, see Swartz, "Conscience Clauses," 284–285.

11. Talcott Parsons, *The Social System* (London: Routledge, 1951), 436.

12. See David J. Rothman, "The Origins and Consequences of Patient Autonomy: A 25-Year Retrospective," *Health Care Analysis* 9, no. 3 (2001): 255, for a more detailed discussion of the patient autonomy movement.

13. Edmund D. Pellegrino, "The Physician's Conscience, Conscience Clauses, and Religious Belief: A Catholic Perspective," *Fordham Urb. L.J.* 30 (2002): 221, 223.

14. Elizabeth Murray, Bernard Lo, Lance Pollack, Karen Donelan, and Ken Lee, "Direct-to-Consumer Advertising: Physicians' Views of Its Effects on Quality of Care and the Doctor-Patient Relationship," *J. Amer. Bd. Fam. Practice* 16 (2003): 513.

15. Erich H. Loewy, "Physicians and Patients: Moral Agency in a Pluralistic World," *J. Med. Human. & Bioethics* 7, no. 1 (1986): 57, 59.

16. R. Alta Charo, "The Celestial Fire of Conscience—Refusing to Deliver Medical Care," *New Eng. J. Med.* 352, no. 24 (2005): 2471, 2472; Martha Minow, "On Being a Religious Professional: The Religious Turn in Professional Ethics," *U. Pa. L. Rev.* 150 (2001): 661, 662–670. Similarly, Lawrence and Curlin have suggested that a major problem in the conscience clause debate is that religious and secular understandings of conscience differ in important ways, which in turn leads to opposing views on what the proper role of conscience ought to be.

See Ryan E. Lawrence and Farr A. Curlin, "Clash of Definitions: Controversies about Conscience in Medicine," *Amer. J. Bioethics* 7, no. 12 (2007): 10.

17. Martin Benjamin, "Conscience," in Stephen G. Post, ed., *The Encyclopedia of Bioethics*, 3rd ed. (New York: Macmillan Reference, 2004), 513.

18. John Rawls, *Political Liberalism* (New York: Columbia University Press, 1993), xxiv–xxvi.

19. Minow, "On Being a Religious Professional," 669.

20. Minow, "On Being a Religious Professional," 665.

21. James A. Monroe, "Morality, Politics, and Health Policy," in David Mechanic, Lynn B. Rogut, David C. Colby, and James R. Knickman, eds., *Policy Changes in Modern Health Care* (New Brunswick, NJ: Rutgers University Press, 2004).

22. For example, Benjamin argues that by definition, conscience is concerned with the consequences to oneself associated with performing an act, not necessarily with the objective or universal wrongness of the act itself. Benjamin, "Conscience," 514. See also Jeffrey Blustein, "Doing What the Patient Orders: Maintaining Integrity in the Doctor-Patient Relationship," *Bioethics* 7, no. 4 (1993): 289, 313.

23. Smearman, "Drawing the Line," 501.

24. For a comparison of what the authors refer to as religious and secular concepts of conscience, see Lawrence and Curlin, "Clash of Definitions."

25. Sabrina Rubin Erdely, "Doctors' Beliefs Hinder Patient Care," *Self Magazine*, June 2007, www.self.com.

26. *N. Coast Women's Care Medical Group v. S.C.* (Benitez), California Supreme Court Case S142892. For more information, see http://appellatecases.courtinfo.ca.gov.

27. *Thomas v. Abdul-Malek*, No. 02-1374 (W.D. Pa July 29, 2004).

28. *Harbeson v. Parke-Davis, Inc.*, 656 P.2d 483 (Wash. 1983) (en banc).

29. *Matthies v. Mastromonaco*, 709 A.2d 238 (N.J. Super. Ct. App. Div. 1998), aff'd, 733 A.2d 456 (N.J. 1999).

30. *Davis v. United States*, 629 F.Supp. 1 (E.D. Ark. 1986).

31. *Walker v. Pierce*, 560 F.2d 609 (4th Cir. 1977).

32. Laura Hunter Dietz, Alan Jacobs, Theresa L. Leming, and John R. Kennel, "Creation and Nature of Relationship," 61 Am. Jur. 2d Physicians, Surgeons, and Other Healers § 130 (2006).

33. See American Medical Association, "Principles of Medical Ethics," § VI; American Medical Association, "Code of Medical Ethics," § 10.05 (2)(b), www.ama-assn.org. However, the AMA eliminates the normally broad discretion to select one's own patients if the physician refuses for discriminatory reasons or in emergency contexts. See also Robert M. Sade, American Medical Association, *Report of the Council on Ethical and Judicial Affairs 6-A-07: Physician Objec-*

tion to Treatment and Individual Patient Discrimination, referred to Committee on Amendments to Constitution and Bylaws (2007), http://www.ama-assn.org/ama1/pub/upload/mm/369/ceja_6a07.pdf ("Principle VI makes clear that physicians may choose whom to serve. Accordingly, except in emergencies, they may refuse to provide a treatment to which they object on the basis of religious or moral beliefs").

34. Farr A. Curlin, Ryan E. Lawrence, Marshall H. Chin, and John D. Lantos, "Religion, Conscience, and Controversial Clinical Practices," *New Eng. J. Med.* 356, no. 6 (2007): 593, 597.

35. C. T. Drechsler, "Liability of Physician Who Abandons Case," 57 A.L.R.2d 432 (2007).

36. 745 Ill. Comp. Stat. 70/4 (2006).

37. See, for example, 745 Ill. Comp. Stat. 70/5 (2006).

38. The following cases dealing with end-of-life conflicts showcase a wide variety of outcomes: *Grace Plaza of Great Neck, Inc. v. Elbaum*, 183 A.D.2d 10 (N.Y. App. Div. 1992) (refusing to force a health-care provider to render treatment contrary to his own conscience, even when no transfer was possible, because a "patient who wishes to abstain from life-saving medical treatment may have the right to do so, but has no right to force a physician to assist, actively or passively, in what the physician himself might regard as the equivalent of suicide"); *In re The Conservatorship of Helga M. Wanglie*, No. PX-91-283 (P. Ct. Minn. Hennepin County June 28, 1991) (retaining husband, who refused to consent to removal of his wife's life-sustaining treatment, as conservator, over physician's objections that the care was futile given the wife's persistent vegetative state); *Gray v. Romeo*, 697 F. Supp. 580, 591 (D.R.I. 1988) (allowing hospital to arrange transfer of the patient seeking to discontinue life-sustaining care, but stating that the hospital must accede to the patient's refusal of artificial nutrition and hydration if the patient could not be "promptly transferred to a health care facility that will respect her wishes"); *Conservatorship of Morrison v. Abramovice*, 253 Cal. Reptr. 530 (Cal. Ct. App. 1988) (refusing to force physicians to remove the patient's feeding tube against their personal moral objections because reasonable transfer was possible); *In re Jobes*, 529 A.2d 434 (N.J. 1987) (compelling a nursing home to continue providing care for a resident seeking removal of her feeding tube because of the nursing home's failure to notify the resident's family of its policy against the withdrawal of artificial nutrition until they requested the tube be removed, as well as because of the difficulty of finding another facility to accept the resident); *In re Requena*, 517 A.2d 886 (N.J. Super. Ct. App. Div. 1986) (holding that a hospital had to continue caring for a patient refusing artificial nutrition and hydration, despite the hospital's "pro-life" policy against discontinuing life-sustaining care and willingness to transfer the patient, because of the burden that such a transfer would impose on the patient who wished to remain in the hospital's care); and *Brophy v. New Eng Sinai Hosp., Inc.*, 497 N.E.2d 626 (Mass. 1986) (refusing to force a hospital to withdraw the patient's artificial feeding tube against its moral and ethical principles, and

instead ordering the hospital to assist in transferring the patient to a suitable facility that would comply with the guardian's decision).

39. See, for example, *Shelton v. Univ. Med. & Dentistry N.J.*, 223 F.3d 220 (3d Cir. 2000) (holding that a hospital had reasonably accommodated the religious beliefs of a nurse opposed to assisting with abortions in compliance with Title VII by offering her a lateral transfer within the hospital, even if this accommodation was not the one most favored by the nurse); *Free v. Holy Cross Hosp.*, 505 N.E.2d 1188, 1190 (Ill. App. Ct. 1987) (interpreting the state conscience clause statute not to cover general ethical beliefs or professional ethics, but rather to protect only sincerely held moral convictions arising from religious beliefs, despite the fact that the statutory definition of conscience included both religious moral convictions and nonreligious beliefs holding a place in the life of the possessor equivalent to that held by religion in the lives of religious adherents); *Warthen v. Toms River Cmty. Mem'l Hosp.*, 488 A.2d 299 (N.J. Super. Ct. App. Div. 1985) (holding that the discharge of a nurse who refused to provide a requested service to a patient on the grounds that it would violate her professional responsibilities according to the American Nurses Association Code for Nurses was not a clear violation of public policy and was therefore permissible); *Kenny v. Ambulatory Centre of Miami, Fla., Inc.*, 400 So.2d 1262 (Fla. Dist. Ct. App. 1981) (adding the undue hardship element of Title VII jurisprudence to the state's conscience clause, despite the statute's lack of exceptions from its prohibitions); *Swanson v. St. John's Lutheran Hosp.*, 597 P.2d 702 (Mont. 1979) (holding that the state's conscience clause, protecting health-care workers from participating in sterilization or abortion procedures against their moral or religious principles, could still apply even if the refusing employee previously participated in the procedure he or she now refused and even if such refusal would place a large burden on the employer). See also Swartz, "Conscience Clauses," 302–314 (discussing individual refusal cases).

40. Federation of State Medical Boards, "Protecting the Public: How State Medical Boards Regulate and Discipline Physicians," www.fsmb.org.

41. Federation of State Medical Boards, *A Guide to the Essentials of a Modern Medical Practice Act*, 10th ed. (Federation of State Medical Boards, 2003): 12–13, www.fsmb.org. State medical acts are very closely based on this model. See, for example, La. Rev. Stat. Ann. § 37:1285 (2006).

42. Norman Daniels, "Duty to Treat or Right to Refuse?," *Hastings Ctr. Rep.* 21, no. 2 (1991): 36.

43. Nora O'Callaghan, "Lessons from Pharaoh and the Hebrew Midwives: Conscientious Objection to State Mandates as a Free Exercise Right," *Creighton L. Rev.* 39 (2006): 561, 567–569.

44. 494 U.S. 872 (1990).

45. Note that this compelling interest test was also in place for laws incidentally burdening the free exercise of religion prior to the Court's decision in *Smith*, which in effect overruled *Sherbert v. Verner*, 374 U.S. 398 (1963). Following

Smith, Congress attempted to reinstate the compelling interest test through the Religious Freedom Restoration Act, which on its face applies to both state and federal law, but was limited to federal application by *City of Boerne v. Flores,* 521 U.S. 507 (1998). See *Stroman v. Lower Merion Twp.,* 2007 U.S. Dist. LEXIS 9408, *9–11 (E.D.Pa. 2007) (briefly explaining the evolution of this area of law).

46. O'Callaghan argues that it likely would. See "Lessons from Pharaoh," 568–570.

47. In fact, see the discussion of potential disincentives in chapter 8 for an argument that the exclusion of potential refusers may not be an appropriate—or even an immediately or always effective—way of improving patient access.

48. *Stormans, Inc. v. Selecky,* W.D.Wa., Case No. C07-5374RBL, Nov. 8, 2007, www.alliancedefensefund.org/UserDocs/StormansPIRuling.pdf.

49. The fundamental problem in this case was not that individuals were allegedly coerced into acting contrary to their religious beliefs, but rather that such coercion was determined to be intentional rather than incidental. *Stormans, Inc.,* at 24.

50. *Catholic Charities of the Diocese of Albany v. Serio,* 7 N.Y.3d 510 (2006). The court's analysis relied heavily on *Employment Division v. Smith.*

51. *Catholic Charities,* 7 N.Y.3d at 527.

52. *St. Agnes Hosp. v. Riddick,* 748 F.Supp. 319, 331 (D. Md. 1990) ("When followers of a particular sect enter into commercial activity as a matter of choice, the limits they accept on their own conduct as a matter of conscience and faith are not to be super-imposed on the statutory schemes which are binding on others in that activity") (quoting *U.S. v. Lee,* 455 U.S. 252, 261 (1982)). *See also Swanner v. Anchorage Equal Rights Commission,* 874 P.2d 274, 283 (Alaska 1994); and *Smith v. Fair Employment & Housing Commission,* 913 P.2d 909 (Cal. 1996).

53. *In re Summers,* 325 U.S. 561 (1945).

54. Dennis Varughese, Note, "Conscience Misbranded! Introducing the Performer v. Facilitator Model for Determining the Suitability of Including Pharmacists within Conscience Clause Legislation," *Temp. L. Rev.* 79 (2006): 649, 681–682.

55. *Jackson v. Metro. Edison Co.,* 419 U.S. 345 (1974).

56. *Moose Lodge No. 107 v. Irvis,* 407 U.S. 163 (1972).

57. A handful of states have never enacted conscience clauses at all, and the protection offered by many existing refusal laws is narrowly limited to objections by particular individuals to particular medical services.

58. Sandeep Jauhar, "A Patient's Demands versus a Doctor's Convictions," *New York Times,* April 3, 2007, www.nytimes.com.

59. This issue is discussed in greater detail in chapter 5.

60. Swartz, "Conscience Clauses," 293.

61. Christopher L. Eisgruber and Lawrence G. Sager, "The Vulnerability of Conscience: The Constitutional Basis for Protecting Religious Conduct," *U. Chi. L. Rev.* 61 (1994): 1245, 1291.

62. Kevin Seamus Hasson, *The Right to Be Wrong: Ending the Culture War over Religion in America* (San Francisco, CA: Encounter Books, 2005), 14.

63. Kent Greenawalt, "Objections in Conscience to Medical Procedures: Does Religion Make a Difference?," *U. Ill. L. Rev.* 2006 (2006): 799, 821. Similarly, cases involving conscientious objection to war have attempted to distinguish conscience from political objections. See, for example, *Gillette v. United States*, 401 U.S. 437, 455–456 (1971).

64. Sylvia A. Law, "Silent No More: Physicians' Legal and Ethical Obligations to Patients Seeking Abortions," *N.Y.U. Rev. L. & Soc. Change* 21 (1995): 279, 303 n. 126. This might also be described as the "dinner party test," referring to whether physicians would feel comfortable telling their friends the types of services they provide to their patients. See Jeremy Laurance, "Abortion Crisis as Doctors Refuse to Perform Surgery," *The Independent*, April 16, 2007, http://news.independent.co.uk.

65. Christopher Meyers and Robert D. Woods, "An Obligation to Provide Abortion Services: What Happens When Physicians Refuse?," *J. Med. Ethics* 22, no. 2 (1996): 115, 118.

66. This sort of objection to abortion seems to be a growing problem, although it could go either way. For example, due to their relative technical ease, abortions may be perceived as "low status and unglamorous," but given the political turmoil surrounding the procedure, doctors who provide it may also be seen as heroic and protecting a worthwhile cause. Laurance, "Abortion Crisis." In fact, some medical students are choosing to make the provision of abortions a significant part of their careers, despite risking isolation, harassment, and hatred, precisely because they feel a moral obligation to do so. They see their choice as an act of defiance against growing restrictions and argue that it is no longer good enough to simply support abortion rights by voting for liberal politicians. Stephanie Simon, "Aspiring Abortion Doctors Drawn to Embattled Field," *LA Times*, May 22, 2007, www.latimes.com.

67. 182 Cal. Rptr. 225 (Cal. Ct. App. 1982).

68. See John Balint, "There Is a Duty to Treat Noncompliant Patients," *Seminars in Dialysis* 14, no. 1 (2001): 28; Alister Browne, Brent Dickenson, and Rena Van Der Waal, "The Ethical Management of the Noncompliant Patient," *Cambridge Q. Healthcare Ethics* 12 (2003): 289.

69. Charo, "The Celestial Fire of Conscience," 2472; Elizabeth Fenton and Loren Lomasky, "Dispensing with Liberty: Conscientious Refusal and the 'Morning-After Pill,'" *J. Med. & Phil.* 30 (2005): 579, 580.

70. For example, devout Roman Catholic physicians may feel themselves bound to the Vatican's most recent ruling that it is wrong to stop administering food and water to patients in a vegetative state even if they will never regain con-

sciousness. John Thavis, "Food, Water Must Be Provided to Vegetative Patients," *Catholic News Service*, September 14, 2007, www.catholic.org.

71. Swartz, "Conscience Clauses," 296.

72. Robert K. Vischer, "Conscience in Context: Pharmacist Rights and the Eroding Moral Marketplace," *Stan. L. & Pol'y Rev.* 17 (2006): 83, 89–97.

73. Compare Mary K. Collins, "Conscience Clauses and Oral Contraceptives: Conscientious Objection or Calculated Obstruction?," *Ann. Health L.* 15 (2006): 37, 38, with Lynn D. Wardle, "Protecting the Rights of Conscience of Health Care Providers," *J. Legal Med.* 14 (1993): 177, 188.

74. See Ronald Dworkin's theory of rights as trumps, in his book *Taking Rights Seriously* (Cambridge: Harvard University Press, 2005).

75. Jeremy Waldron, *Law and Disagreement* (Oxford: Oxford University Press, 1999), 12 (quoting Thomas Hobbes, *Leviathan* (1651), 33).

76. Vischer, "Conscience in Context," 96.

77. Mark A. Hall, Mary Anne Bobinski, and David Orentlicher, *Health Care Law and Ethics*, 6th ed. (New York: Aspen Publishers, 2003), 117; Kenneth R. Wing, "The Right to Health Care in the United States," *Ann. Health L.* 2 (1993): 161.

78. See *Payton v. Weaver*, 182 Cal. Rptr. 225 (Cal. Ct. App. 1982).

79. *Planned Parenthood v. Casey*, 505 U.S. 833, 876 (1992).

80. Wardle, "Protecting the Rights of Conscience," 211. See, for example, *Beal v. Doe*, 432 U.S. 438 (1977) (holding that the omission of nontherapeutic abortions from Medicaid funding is not unduly burdensome, and thus violates no constitutional right); *Maher v. Roe*, 432 U.S. 464 (1977) (holding that the provision of Medicaid funds for childbirth but not abortion is not discriminatory against any suspect class, and that this policy infringes no fundamental rights because a woman seeking an abortion is not disadvantaged by the state's choice).

81. *Griswold v. Conn.*, 381 U.S. 479 (1965).

82. *Cruzan v. Director, Missouri Dep't of Health*, 497 U.S. 261, 270 (1990).

83. "We do not recognize any right to force a health provider to render treatment which is contrary to his or her own conscience." *Grace Plaza of Great Neck, Inc. v. Elbaum*, 588 N.Y.S.2d 853, 859 (N.Y. App. Div. 1992).

84. 42 U.S.C. § 1395dd (1986).

85. See Melissa K. Stull, "Construction and Application of Emergency Medical Treatment and Active Labor Act," 104 A.L.R. Fed. 166, §§ 8(a) and (b) (2006).

86. Laura Hunter Dietz, Alan J. Jacobs, Theresa L. Leming, Lucas Martin, Jeffrey Shampo, Eric Surette, and Lisa Zakolski, "Moral or Humanitarian Considerations; Duty to Aid or Protect Others," 57A Am. Jur. 2d Negligence § 90 (2006).

87. "In the absence of statute providing otherwise, a physician or surgeon is under no legal obligation to render professional services to everyone who applies to

him, not even in the case of an emergency." Laura Hunter Dietz, Alan Jacobs, Theresa L. Leming, and John R. Kennel, "Obligation to Practice or Accept Professional Employment," 61 Am Jur 2d Physicians, Surgeons, and Other Healers § 121 (2007).

88. See, for example, AMA, "Principles of Medical Ethics," § VI; 745 Ill. Comp. Stat. 70/6. (2006); Okla. Stat. Ann. Tit. 63 § 1–741 (B) (West 2005); and Nev. Rev. Stat. Ann. §§ 449.191 and 632.475.

89. *Schloendorff v. Soc'y of N.Y. Hosp.*, 211 N.Y. 125, 129 (1914).

90. See, for example, N.Y. Pub. Health Law § 2805-d (McKinney 2006).

91. Laura Hunter Dietz, Alan Jacobs, Theresa L. Leming, and John R. Kennel, "Duty to Refer Patient to Specialist or Qualified Practitioner," 61 Am. Jur. 2d Physicians, Surgeons, and Other Healers § 214 (2006).

92. Laura Hunter Dietz, Alan Jacobs, Theresa L. Leming, and John R. Kennel, "Obligation to Practice or Accept Professional Employment," 61 Am. Jur. 2d Physicians, Surgeons, and Other Healers § 121 (2006).

93. *Ricks v. Budge*, 64 P.2d 208, 211–212 (Utah 1937). See also Laura Hunter Dietz, Alan J. Jacobs, Theresa L. Leming, and John R. Kennel, 61 Am. Jur. 2d Physicians, Surgeons, and Other Healers §§ 217–218 (2006).

94. 42 U.S.C. § 2000(e).

95. *Trans World Airlines, Inc. v. Hardison*, 432 U.S. 63, 84 (1977). For further discussion of Title VII, see chapter 3.

96. Federation of State Medical Boards, *Guide*, 2.

97. See, for example, Howard Brody and Susan S. Night, "The Pharmacist's Personal and Professional Integrity," *Am. J. Bioethics* 7, no. 6 (2007): 16.

Chapter 2

1. Louise Arnold and David Thomas Stern, "What Is Medical Professionalism?," in David Stern, ed., *Measuring Medical Professionalism* (Oxford: Oxford University Press, 2006), 15.

2. Herbert M. Swick, "Toward a Normative Definition of Medical Professionalism," *Acad. Med.* 75, no. 6 (2000): 612, 614–615.

3. Norman Daniels, *Just Health: A Population View* (Cambridge: Cambridge University Press, 2007), 246. But note that even the conscientious refuser could make an argument that he is completely subordinating his own interests to the patient's if that refuser believed that the patient's request was against that patient's own interest—perhaps that she would regret having an abortion, for example—*and* the refuser would suffer some penalty for refusing to satisfy it.

4. American Medical Association, "Code of Medical Ethics," § 10.015.

5. American Medical Association, "Code of Medical Ethics," § 10.05. Attorneys are also, as a matter of professional ethics, generally free to decline their services

to potential clients. See, for example, New York Lawyer's Code of Professional Responsibility, Ethical Consideration 2–26.

6. Martha Swartz, "Conscience Clauses or Unconscionable Clauses: Personal Beliefs versus Professional Responsibilities," *Yale J. Health Pol'y, L., & Ethics* 6 (2006): 269, 342.

7. U.S. Census Bureau, "Income, Poverty, and Health Insurance Coverage in the United States: 2006 Report," www.census.gov.

8. R. Alta Charo, "The Celestial Fire of Conscience—Refusing to Deliver Medical Care," *New Eng. J. Med.* 352, no. 24 (2005): 2471, 2472; Edmund D. Pellegrino, "The Physician's Conscience, Conscience Clauses, and Religious Belief: A Catholic Perspective," *Fordham Urb. L.J.* 30 (2002): 221, 229.

9. Louis-Jacques Van Bogaert, "The Limits on Conscientious Objection to Abortion in the Developing World," *Developing World Bioethics* 2, no. 2 (2002): 131, 136.

10. Charo, "The Celestial Fire of Conscience," 2472. See also Walter Robinson, "In the Line of Duty: SARS and Physician Responsibility in Epidemics," *Lahey Clinic Med. Ethics* (Spring 2006): 5, 7; Norman Daniels, "Duty to Treat or Right to Refuse?," *Hastings Ctr. Rep.* 21, no. 2 (1991): 36.

11. B. M. Dickens and R. J. Cook, "The Scope and Limits of Conscientious Objection," *Int'l J. Gynecology & Obstetrics* 71, no. 1 (2000): 71, 72.

12. Robert M. Veatch, "Doctor Does Not Know Best: Why in the New Century Physicians Must Stop Trying to Benefit Patients," *J. Med. & Phil.* 25, no. 6 (2000): 701, 705. See also Robert M. Veatch, "Modern vs. Contemporary Medicine: The Patient-Provider Relation in the Twenty-First Century," *Kennedy Instit. Ethics J.* 6, no. 4 (1996): 366, 368.

13. A. Torda, "How Far Does A Doctor's 'Duty of Care' Go?," *Internal Med. J.* 35, no. 5 (2005): 295, 295.

14. R. Smith, "Medicine's Core Values," *Brit. Med. J.* 309 (1994): 1247, 1247. See also John K. Inglehart, "Grassroots Activism and the Pursuit of an Expanded Physician Supply," *New Eng. J. Med.* 358, no. 16 (2008): 1741, 1742–1743 (Table 1).

15. Denise M. Dudzinski, "Integrity in the Relationship between Medical Ethics and Professionalism," *Am. J. Bioethics* 4, no. 2 (2004): 26, 27.

16. Edmund D. Pellegrino, "Character, Virtue, and Self-Interest in the Ethics of the Professions," *J. Contemp. Health L. & Pol'y* 5 (1989): 53, 57. See also Edmund D. Pellegrino, "Professionalism, Profession and the Virtues of the Good Physician," *Mt. Sinai J. Med.* 69, no. 6 (2002): 378.

17. Daniels, *Just Health*, 249–252; Daniels, "Duty to Treat."

18. Norman Daniels, personal communication, October 19, 2006.

19. Rebecca Dresser, "Professionals, Conformity, and Conscience," *Hastings Ctr. Rep.* 35, no. 6 (2005): 9, 9.

20. Daniels, *Just Health*, 252–253.

21. See David T. Ozar, "Profession and Professional Ethics," in Stephen G. Post, ed., *The Encyclopedia of Bioethics*, 3rd ed. (New York: Macmillan Reference, 2004), 2167; Kenneth A. Richman, "Pharmacists and the Social Contract," *Am. J. Bioethics* 7, no. 6 (2007): 15, 15.

22. Ozar, "Profession and Professional Ethics," 2167.

23. Daniels, *Just Health*, 248.

24. Daniel Sokol, "Virulent Epidemics and Scope of Healthcare Workers' Duty of Care," *Emerging Infectious Diseases* 12, no. 8 (2006): 1238, 1238.

25. Robinson, "In the Line of Duty," 7.

26. Daniels, "Duty to Treat."

27. Daniels, *Just Health*, 248–255.

28. Ryan E. Lawrence and Farr A. Curlin, "Clash of Definitions: Controversies about Conscience in Medicine," *Amer. J. Bioethics* 7, no. 12 (2007): 10; Christopher L. Eisgruber and Lawrence G. Sager, "The Vulnerability of Conscience: The Constitutional Basis for Protecting Religious Conduct," *U. Chi. L. Rev.* 61 (1994): 1245, 1262; Pellegrino, "The Physician's Conscience, Conscience Clauses, and Religious Belief," 231, 239.

29. Dresser, "Professionals, Conformity, and Conscience," 9.

30. Martha Swartz, "Health Care Providers' Rights to Refuse to Provide Treatment on the Basis of Moral or Religious Beliefs," *Health Lawyer* 19 (2006): 25, 28.

31. Richard L. Cruess, Sylvia R. Cruess, and Sharon E. Johnston, "Professionalism and Medicine's Social Contract," *J. Bone & Joint Surg.* 82-A (2000): 1189, 1192.

32. See AMA, "Code of Medical Ethics," § 10.05 (3)(c). However, for a critique of professional associations supporting conscientious refusal, see Inmaculada de Melo-Martin, "Should Professional Associations Sanction Conscientious Refusals?," *Am. J. Bioethics* 7, no. 6 (2007): 23.

33. "Health and Ethics Policies of the AMA House of Delegates," § 5.990. The AMA has also stated that "every physician remains free to decide whether to participate in stem cell research or to use its products." AMA, "Code of Medical Ethics," § 2.146.

34. American College of Obstetricians and Gynecologists (ACOG), *Ethics in Obstetrics and Gynecology*, 2nd ed. (Washington, D.C.: American College of Obstetricians and Gynecologists, 2004), 16, www.acog.org.

35. In 2007, ACOG issued a committee opinion stating, among other things, that conscientious refusers must nonetheless fully inform patients of their reproductive options, provide referrals to willing providers, and even provide the objectionable service under certain circumstances. ACOG, Committee Opinion, "The Limits of Conscientious Refusal in Reproductive Medicine," No. 385 (November 2007), www.acog.org. As a result of the controversy surrounding this opinion, ACOG clarified that it was just that—an opinion, not a formal part of

the group's official Code of Ethics that board certified ob-gyns must obey. Julie Rovner, "New Ob/Gyn Guidelines Stir Ethics, Legal Debate," *NPR* (March 19, 2008), www.npr.org. American College of Obstetricians and Gynecologists, Executive Board, "ACOG Statement of Policy: Abortion Policy" (Washington, D.C: American College of Obstetricians and Gynecologists, 2004).

36. Christopher Meyers and Robert D. Woods, "An Obligation to Provide Abortion Services: What Happens When Physicians Refuse?," *J. Med. Ethics* 22, no. 2 (1996): 115, 115.

37. Karen E. Adams, "Moral Diversity among Physicians and Conscientious Refusal of Care in the Provision of Abortion Services," *J. Am. Med. Women's Ass'n* 58, no. 4 (2003): 223, 224. The price of training was taken from Dennis Cauchon, "Medical Miscalculation Creates Doctor Shortage," *USA Today*, March 2, 2005, www.usatoday.com.

38. Eisgruber and Sager, "Vulnerability," 1263.

39. See Daniels, *Just Health*, 214–244.

40. Julian Savulescu, "Conscientious Objection in Medicine," *Brit. Med. J.* 332 (2006): 294. This article stirred significant controversy; in fact, out of fifty-eight online responses, most of which were written by practicing physicians, Rosamond Rhodes offered the sole supportive comment, and she is a PhD. Her perspective is discussed in more detail later in this section.

41. Savulescu, "Conscientious Objection," 295. Note that while legality and patient desires are often unambiguous, the other two criteria listed by Savulescu here—that the services are beneficial to the patient and part of a just health-care system—are open to an extraordinary amount of debate. Is physician assisted suicide or abortion beneficial to the patient? Clearly there will be disagreement on this score, which is partly why these criteria cannot be adopted wholesale as the standard for establishing the obligations of state licensing boards, discussed in chapter 5. Savulescu also notes that legality itself may sometimes be in question "Conscientious Objection," 296.

42. Savulescu, "Conscientious Objection," 296.

43. Savulescu, "Conscientious Objection," 294.

44. Savulescu, "Conscientious Objection," 296.

45. *Rodriguez v. City of Chicago*, 156 F.3d 771 (7th Cir. 1998).

46. *Shelton*, 223 F.3d 220, 228 (3d Cir. 2000).

47. *Endres*, 349 F.3d 922 (7th Cir. 2003), cert. denied, 124 S.Ct. 2032 (2004).

48. R. Alta Charo, "Health Care Provider Refusals to Treat, Prescribe, Refer, or Inform: Professionalism and Conscience" (American Constitution Society, February 2007), 13, www.americanconstitutionsociety.org.

49. However, we will see in chapter 6 that even if others are willing to provide the service, some refusals are still harmful, such as those that are the product of invidious discrimination, rendering them impermissible.

50. These issues are discussed in greater detail in chapters 8 and 9.

51. Savulescu, "Conscientious Objection," 294.

52. Savulescu, "Conscientious Objection," 294.

53. Jed Miller, Note, "The Unconscionability of Conscience Clauses: Pharmacists and Women's Access to Contraception," *Health Matrix* 16 (2006): 237, 265.

54. The traditionalist and dominant, although not exclusive, view of the role of an attorney is to act as amoral counsel—a "hired gun" serving whatever goals the client selects for herself, though within the confines of the law and professional ethics, ambiguities in which should always be interpreted in favor of permitting the client's goal. According to this model, a lawyer's choice to refuse professional services on the basis of personal moral reasons is considered self-indulgent, unacceptably preferring the lawyer's own interests to her professional duties. Katherine R. Kruse, "Lawyers, Justice, and the Challenge of Moral Pluralism," *Minn. L. Rev.* 90 (2005): 389, 390. See also Russell G. Pearce, "The Legal Profession as a Blue State: Reflections on Public Philosophy, Jurisprudence, and Legal Ethics," *Fordham L. Rev.* 75 (2007): 1339, 1341, 1361–1362.

55. For example, while 50 percent of the public reports being extremely or very confident in the medical profession, only 19 percent say the same about the legal profession. See American Bar Association, "Public Perceptions of Lawyers: Consumer Research Findings" (2002), www.abanet.org.

56. Ezekiel J. Emanuel and Linda L. Emanuel, "Four Models of the Physician-Patient Relationship," *J. Amer. Med. Ass'n* 267, no. 16 (1992): 2221, 2224.

57. Savulescu, "Conscientious Objection," 296.

58. Rosamond Rhodes, "The Ethical Standard of Care," *Am. J. Bioethics* 6, no. 2 (2006): 76, 78.

59. Rosamond Rhodes, "The Priority of Professional Ethics Over Personal Morality," February 9, 2006, BMJ Online Rapid Responses to Savulescu's "Conscientious Objection," www.bmj.com.

60. Rhodes, "Standard of Care," 78. See also Swartz, "Conscience Clauses," 339.

61. Rhodes, "Priority."

62. But note that the Board of Professional Responsibility of the Supreme Court of Tennessee required an attorney morally and religiously opposed to abortion to represent a minor seeking to obtain that procedure without parental consent. Formal Op. 96-F-140 (1996). The opinion of the board portrayed lawyers as "merely the mouthpiece through which the clients' claims are presented to the court." Teresa Stanton Collett, "Professional versus Moral Duty: Accepting Appointments in Unjust Civil Cases," *Wake Forest L. Rev.* 32 (1997): 635, 645.

63. Jeffrey Blustein and Alan R. Fleischman, "The Pro-Life Maternal-Fetal Medicine Physician: A Problem of Integrity," *Hastings Ctr. Rep.* 25, no. 1 (1995): 22.

64. ACOG Executive Board, "ACOG Statement of Policy: Abortion Policy."

65. Blustein and Fleischman, "Pro-Life Maternal-Fetal Medicine Physician."

66. Consider a baseball analogy: a player may be an excellent pitcher who contributes tremendously to the success of his team, but he may also be a terrible hitter. Rather than losing out on his talent, the team may select a designated hitter to substitute for the pitcher, who would otherwise have to be placed in the batting lineup. The pitcher who does not hit is not shirking any responsibilities owed to his team and is a valuable commodity. Similarly, a physician who refuses to provide some medical service may nonetheless be an important addition to the profession, so long as there is the equivalent of a designated hitter willing to provide that service in his stead. Of course, even the designated-hitter rule has experienced its fair share of controversy. See, for example, Christopher Caldwell, "A DHumb Idea at 25," *The Weekly Standard*, April 13, 1998.

67. Lawrence and Curlin, "Clash of Definitions."

68. Sean Murphy, "Freedom of Conscience and the Needs of the Patient," presentation at the 2001 Obstetrics and Gynecology Conference: New Developments–New Boundaries, Banff, Alberta, 4, www.consciencelaws.org.

69. Larry Cunningham, "Can a Catholic Lawyer Represent a Minor Seeking a Judicial Bypass for an Abortion? A Moral and Canon Law Analysis," *J. Cath. Legal Stud.* 44, no. 2 (2005): 379, 380. See also Piers Benn, "The Role of Conscience in Medical Ethics," in Nafiska Athanssoulis, ed., *Philosophical Reflections on Medical Ethics* (New York: Palgrave Macmillan, 2005), 171–174; Jeffrey Blustein, "Doing What the Patient Orders: Maintaining Integrity in the Doctor-Patient Relationship," *Bioethics* 7, no. 4 (1993): 289, 313; Collett, "Professional versus Moral Duty," 649.

70. Lynn A. Jansen, "No Safe Harbor: The Principle of Complicity and the Practice of Voluntary Stopping of Eating and Drinking," *J. Med. & Phil.* 29, no. 1 (2004): 61, 65.

71. Edmund D. Pellegrino, "Commentary: Value Neutrality, Moral Integrity, and the Physician," *J. L. Med. & Ethics* 28, no. 1 (2000): 78, 79. See also Farr A. Curlin and Daniel E. Hall, "Strangers or Friends? A Proposal for a New Spirituality-in-Medicine Ethic," *J. Gen. Intern. Med.* 20, no. 4 (2005): 370, 373; Timothy Floyd, "Does Professionalism Leave Room for Religious Commitment?," *Fordham Urb. L.J.* 26 (1999): 875, 883.

72. Pellegrino, "Commentary," 78–79.

73. J. Budziszewski, "Handling Issues of Conscience," 1999 Beatty Memorial Lecture, McGill University College of Education, www.consciencelaws.org.

74. Edmund D. Pellegrino, "Patient and Physician Autonomy: Conflicting Rights and Obligations in the Physician-Patient Relationship," *J. Contemp. Health L. & Pol'y* 10 (1993): 47, 58–59.

75. Pellegrino, "Patient and Physician Autonomy," 48–49.

76. Pellegrino, "Patient and Physician Autonomy," 48–52; Pellegrino, "Commentary," 80; Thomas May, "Rights of Conscience in Health Care," *Soc. Theory & Practice* 27, no. 1 (2001): 111, 111.

77. Pellegrino, "Character, Virtue, and Self-Interest," 57; Pellegrino, "Patient and Physician Autonomy," 52.

78. Charles Fried, "The Lawyer as Friend: The Moral Foundations of the Lawyer-Client Relation," *Yale L.J.* 85 (1976): 1060, 1078–1079 (emphasis added). See also Teresa Stanton Collett, "The Common Good and the Duty to Represent: Must the Last Lawyer in Town Take Any Case?," *S. Tex. L. Rev.* 40 (1999): 137, 160.

79. Pellegrino, "The Physician's Conscience, Conscience Clauses, and Religious Belief," 243; Pellegrino, "Patient and Physician Autonomy," 64; and Pellegrino, "Commentary," 80.

80. Rhodes, "Priority."

81. Fried, "Lawyer as Friend," 1078–1079; Martha Minow, "On Being a Religious Professional: The Religious Turn in Professional Ethics," *U. Pa. L. Rev.* 150 (2001): 661, 682.

82. Pellegrino, "The Physician's Conscience, Conscience Clauses, and Religious Belief," 242.

83. Jeremy Laurance, "Abortion Crisis as Doctors Refuse to Perform Surgery," *The Independent*, April 16, 2007, www.independent.co.uk.

84. Pellegrino, "The Physician's Conscience, Conscience Clauses, and Religious Belief," 231.

85. Robert E. Goodin, *Protecting the Vulnerable: A Reanalysis of Our Social Responsibilities* (Chicago: University of Chicago Press, 1985), 33–35.

86. Ozar, "Profession and Professional Ethics," 2162–2163.

87. Ozar, "Profession and Professional Ethics," 2162–2163. See also Pearce, "The Legal Profession as a Blue State," 1356; Sylvia R. Cruess, Sharon Johnston, and Richard L. Cruess, "Professionalism for Medicine: Opportunities and Obligations," *Iowa Orthopaedic J.* 24 (2004): 9, 9.

88. For example, Washington State's rule requiring pharmacies to satisfy patient prescriptions regardless of conscientious objections nonetheless provides that pharmacies are under no obligation "to deliver a drug or device without payment of their usual and customary or contracted charge." Wash. Admin. Code § 246-869-010.

89. Goodin states that "the vulnerability of people in need of professional help but unable to obtain it explains the obligation imposed upon both doctors and lawyers, collectively and to some extent individually, to provide services free or at reduced rates to those who cannot afford the ordinary fees." Goodin, *Protecting the Vulnerable*, 68.

90. Daniels, *Just Health*, 258–260.

91. Charo, "The Celestial Fire of Conscience," 2473. See also Charo, "Health Care Provider Refusals to Treat, Prescribe, Refer, or Inform," 9, 15–17.

92. Rebecca S. Dresser, "Freedom of Conscience, Professional Responsibility, and Access to Abortion," in Belinda Bennett, ed., *Abortion* (Burlington, VT: Ashgate, 2004), 455.

93. Daniels, *Just Health*, 267 n. 4.

94. Goodin, *Protecting the Vulnerable*, 41.

95. Goodin, *Protecting the Vulnerable*, 119–122.

96. Gregory L. Weiss and Lynne E. Lonnquist, *The Sociology of Health, Healing, and Illness*, 5th ed. (Upper Saddle River, NJ: Prentice Hall, 2005), 281.

97. Charo, "The Celestial Fire of Conscience," 2473. See also Swartz, "Conscience Clauses," 279; Meyers and Woods, "Obligation to Provide Abortion Services," 117.

98. Charo, "The Celestial Fire of Conscience," 2473.

Chapter 3

1. Sylvia A. Law, "Silent No More: Physicians' Legal and Ethical Obligations to Patients Seeking Abortions," *N.Y.U. Rev. L. & Soc. Change* 21 (1995): 279, 315 (citing Vt. Stat. Ann. tit. 13, § 101 (1974)).

2. Iain T. Benson, "'Autonomy,' 'Justice,' and the Legal Requirement to Accommodate the Conscience and Religious Beliefs of Professionals in Health Care," 2001, 4, www.consciencelaws.org.

3. J. David Bleich, "The Physician as a Conscientious Objector," *Fordham Urb. L.J.* 30 (2002): 245, 245.

4. Sanford Levinson, "Identifying the Jewish Lawyer: Reflections on the Construction of Professional Identity," *Cardozo L. Rev.* 14 (1993): 1577, 1586.

5. Howard Lesnick, "The Religious Lawyer in a Pluralist Society," *Fordham L. Rev.* 66 (1998): 1469, 1489–1490.

6. Katherine Dowling, "Prolife Doctors Should Have Choices, Too," *U.S. Catholic*, March 2001, 27.

7. See James H. Jones, *Bad Blood: The Tuskegee Syphilis Experiment* (New York: Free Press, 1993).

8. See Scott Krugman, "The Willowbrook Hepatitis Studies Revisited: Ethical Aspects," *Rev. Infectious Diseases* 8, no. 1 (1986): 157–162.

9. See Robert Jay Lifton, *The Nazi Doctors: Medical Killing and the Psychology of Genocide* (New York: Basic Books, 2000).

10. See Richard S. Matthews, "Indecent Medicine: In Defense of the Absolute Prohibition against Physician Participation in Torture," *Am. J. Bioethics* 6, no. 3 (2006): W34.

11. See Peter A. Clark, "Physician Participation in Executions: Care Giver or Executioner?," *J. L. Med. & Ethics* 34, no. 1 (2006): 95.

12. John Stuart Mill, "On Liberty," in *The Basic Writings of John Stuart Mill* (New York: Modern Library, 2002), 106–107.

13. Mill, "On Liberty," 78.

14. S.D.C.L. § 34-23A-10.1 (2005).

15. *Planned Parenthood Minn. v. Rounds*, 467 F.3d 716 (8th Cir. 2006), rehearing en banc, No. 05-3093 (8th Cir. Argued Apr. 11, 2007).

16. Louis-Jacques Van Bogaert, "The Limits on Conscientious Objection to Abortion in the Developing World," *Developing World Bioethics* 2, no. 2 (2002): 131, 134–135; Edmund D. Pellegrino, "Commentary: Value Neutrality, Moral Integrity, and the Physician," *J. L. Med. & Ethics* 28, no. 1 (2000): 78, 79.

17. It is essential to recognize that, consent aside, legality and morality are two different things, and morality is not necessarily defined by the democratic process or even by the Constitution—slavery is the perfect example of this fact. We hope that the two overlap in the most important cases, but we cannot rely on the law to mirror morality in every circumstance; "human law is fallible and therefore may be rightly ignored in some circumstances." Ryan E. Lawrence and Farr A. Curlin, "Clash of Definitions: Controversies about Conscience in Medicine," *Amer. J. Bioethics* 7, no. 12 (2007): 10.

18. Mark R. Wicclair, "Conscientious Objection in Medicine," *Bioethics* 14, no. 3 (2000): 205, 210.

19. Mill, "On Liberty," 19.

20. Frances L. Key, "Female Circumcision/Female Genital Mutilation in the United States: Legislation and Its Implications for Health Providers," *J. Am. Med. Women's Ass'n* 52, no. 4 (1997): 179, 179.

21. Compare American College of Obstetricians and Gynecologists, Committee on Gynecologic Practice and Committee on International Affairs, "Female Genital Mutilation," *Int. J. Gynecology & Obstetrics* 49, no. 2 (1995): 209, 209 ("The American College of Obstetricians and Gynecologists joins many other major organizations (World Health Organization, United Nations International Children's Emergency Fund, International Federation of Gynecology and Obstetrics, and the American Medical Association) in opposing all forms of medically unnecessary surgical modification of the female genitalia"), with Julie D. Cantor, "When an Adult Female Seeks Ritual Genital Alteration: Ethics, Law, and the Parameters of Participation," *Plastic & Reconstructive Surgery* 117, no. 4 (2006): 1158, 1163 ("Americans alter their bodies in profound and controversial ways, from tattooing and piercing to miscellaneous implants and gender reassignment. These modifications involve personal assessments of need, risk, and benefit. Against that backdrop, arguments to deny voluntary genital alteration to women ring hollow.... For some patients, the benefits [of female circumcision] will be quite clear, and adults who conclude that such surgery is right for them should not be flatly overruled by an American society that may not endorse or understand their cultural preferences").

22. Stephen Wear, Susan Lagaipa, and Gerald Logue, "Toleration of Moral Diversity and the Conscientious Refusal by Physicians to Withdraw Life-Sustaining Treatment," *J. Med. & Phil.* 19, no. 2 (1994): 147, 156. There is no constitutional right to physician-assisted suicide or voluntary euthanasia, but there is nothing to prevent states from legalizing either practice.

23. Patrick McIlheran, "Don't Divorce Job from Conviction in Health Care," *Milwaukee Journal Sentinel*, February 13, 2005.

24. See, for example, Kent Greenawalt, "Objections in Conscience to Medical Procedures: Does Religion Make a Difference?," *U. Ill. L. Rev.* (2006): 799, 821 n. 88.

25. Courtney Miller, Note, "Reflections on Protecting Conscience for Health Care Providers: A Call for More Inclusive Statutory Protections in Light of Constitutional Considerations," *S. Cal. Rev. L. & Soc. Just.* 15 (2006): 327, 356. See also Wicclair, "Conscientious Objection in Medicine," 207.

26. Miller, "Reflections," 357.

27. Mill, "On Liberty," 80 (emphasis added).

28. W. Bradley Wendel, "Institutional and Individual Justification in Legal Ethics: The Problem of Client Selection," *Hofstra L. Rev.* 34 (2006): 987, 992.

29. Robert K. Vischer, "Conscience in Context: Pharmacist Rights and the Eroding Moral Marketplace," *Stan. L. & Pol'y Rev.* 17 (2006): 83, 117. See also Robert Vischer, "The Pharmacists Wars," *American Enterprise Online*, February 14, 2006, www.taemag.com.

30. In light of evidence that physicians often make treatment decisions that reflect their own preferences rather than the preferences of their patients, as well as the fact that it may not always be possible for physicians to discern their patients' preferences, Orentlicher explains that if "patients choose physicians whose values coincide with their own, it is much more likely that treatment decisions will coincide with the patient's values and preferences." David Orentlicher, "Trends in Health Care Decisionmaking: The Limits of Legislation," *Md. L. Rev.* 53 (1994): 1255, 1304.

31. Robert M. Veatch, *A Theory of Medical Ethics* (New York: Basic Books, 1981), 136.

32. Robert M. Veatch, *The Patient-Physician Relation: The Patient as Partner, Part 2* (Bloomington: Indiana University Press, 1991), 153.

33. Veatch, *The Patient-Physician Relation*, 31.

34. Veatch, *A Theory of Medical Ethics*, 247. Note that other commentators have advocated even more individual negotiation between doctors and their patients with hardly any constraints. For example, John Alexander suggests that the role of a professional is the competent provision of those services that are a "normal part of the services that the professionals would be called upon to provide in the normal course of their professional lives within the specific practice with which they are associated." While physicians would have no right to refuse

to provide services they should have expected their patients to ask for, it would be within their power to describe and circumscribe their own "normal course" by choosing medical specialties that generally conform to their desires, beliefs, and preferences, and more importantly, by carefully selecting the type of *practice* in which they will engage. Once a practice is selected, the physician would be bound by the associated promises, but they are the sole check on his or her ability to conscientiously refuse. See John K. Alexander, "Promising, Professional Obligations, and the Refusal to Provide Service," *HEC Forum* 17, no. 3 (2005): 178. This is clearly not the standard model of medical professionalism, and adopting it could destroy the very notion of a profession.

35. Veatch, *A Theory of Medical Ethics*, 198.

36. Veatch, *Patient-Physician Relation*, 153.

37. Arthur Applbaum, "Doctor, Schmocter: Practice Positivism and Its Complications," in Robert B. Baker, Arthur L. Caplan, Linda L. Emanuel, and Stephen R. Latham, eds., *The American Medical Ethics Revolution* (Baltimore: Johns Hopkins University Press, 1999), 144–156.

38. David Mechanic, "In My Chosen Doctor I Trust," *Brit. Med. J.* 329 (2004): 1418, 1418.

39. Mechanic, "In My Chosen Doctor I Trust," 1418.

40. Farr A. Curlin and Daniel E. Hall, "Strangers or Friends? A Proposal for a New Spirituality-in-Medicine Ethic," *J. Gen. Intern. Med.* 20 (2005): 370, 372.

41. Norman Daniels, personal communication, October 17, 2006.

42. Pellegrino, "Commentary," 80.

43. To make this situation as analogous as possible, we must assume that these charges did not arise from any medical behavior on the part of the physician, such as ending the life of a comatose patient by administering a lethal dose of medication. Otherwise, the source of distrust could be that the patient now worries that the physician will do the same to him or her.

44. See, for example, Somnath Saha, Sara H. Taggart, Miriam Komaromy, and Andrew B. Bindman, "Do Patients Choose Physicians of Their Own Race?," *Health Affairs* 19, no. 4 (2000): 76, 82. See also Bradley Gray and Jeffrey Stoddard, "Patient-Physician Pairing: Does Racial and Ethnic Congruity Influence Selection of a Regular Physician?," *J. Cmty. Health* 22, no. 4 (1997): 247, 258; Mary Hegarty Nowlan, "Women Doctors, Their Ranks Growing, Transform Medicine," *Boston Globe*, October 2, 2006, www.boston.com.

45. For a discussion of this issue, see James D. Capozzi and Rosamond Rhodes, "Coping with Racism in a Patient," *J. Bone & Joint Surgery* 88, no. 11 (2006): 2543. Even here, however, there are some open questions. Should a woman's preference for a female gynecologist because she feels more comfortable discussing intimate sexual issues with a person of her own gender play a role in shaping physician supply? Is the Hmong immigrant who prefers a Hmong doctor exhibiting racism?

46. Mill, "On Liberty," 54.

47. See, for example, Miss. Code Ann. § 41-107-5. The Washington and Illinois statutes are similarly broad.

48. Compare *Kenny v. Ambulatory Centre of Miami, Fla., Inc.*, 400 So.2d 1262 (Fla. Dist. Ct. App. 1981) with *Swanson v. St. John's Lutheran Hosp.*, 597 P.2d 702 (Mont. 1979). In *Kenny*, the court added an undue hardship element to the state's conscience clause, thus protecting employers, even though the statute lacked any explicit exemptions from its broad protections of conscientious refusers. In *Swanson*, on the other hand, the court held that the state's conscience clause applied even if the employee's refusal would place a large burden on her employer.

49. Title VII of the Civil Rights Act of 1964, 42 U.S.C. §§ 2000e(j) and 2000e-2(e). Undue hardship is not defined by Title VII; the de minimis standard was introduced by the Supreme Court in *Trans World Airlines, Inc. v. Hardison*, 432 U.S. 63 (1977). Note that Title VII does not preempt state laws that offer greater protection from employment discrimination. See Kenneth W. Biedzynski, Laura Hunter Dietz, Rosemary Gregor, Alan J. Jacobs, Theresa Leming, Bill Lindsley, Lucas Martin, Anne Payne, Jeffrey Shampo, and Eric C. Surette, 45A *Am Jur 2d Job Discrimination* § 2 (2006).

50. The Americans with Disabilities Act, 42 U.S.C. §§ 12111(10) and 12111(8).

51. For an extension of this argument that many existing conscience clauses improperly burden employers, see James A. Sonne, "Firing Thoreau: Conscience and At-Will Employment," *U. Pa. J. Lab. & Emp. L.* 9 (2007): 235.

52. Note, however, that this is an important difficulty facing nonphysician refusers, who may lack the freedom to create their own practice in line with their moral beliefs, but instead have to rely on the willingness of some employer to accommodate those beliefs. This is one aspect of the conscience clause debate that makes it somewhat simpler to resolve for doctors than for other types of refusers.

53. Vischer, "Conscience in Context," 119.

54. Rob Stein, "Medical Practices Blend Health and Faith: Doctors, Patients Distance Themselves from Care They Consider Immoral," *Washington Post*, August 31, 2006, A1.

55. Stein, "Medical Practices Blend Health and Faith," A1.

56. Rob Stein, "Institute Practices Reproductive Medicine—and Catholicism," *Washington Post*, October 31, 2006, A14.

57. Mary K. Collins, "Conscience Clauses and Oral Contraceptives: Conscientious Objection or Calculated Obstruction?," *Annals Health L.* 15 (2006): 37, 57.

58. Stein, "Medical Practices Blend Health and Faith."

59. James F. Childress and Mark Siegler, "Metaphors and Models of Doctor-Patient Relationships: Their Implications for Autonomy," *Theor. Med.* 5, no. 1 (1984): 17, 20.

60. Martha Swartz, "Conscience Clauses or Unconscionable Clauses: Personal Beliefs versus Professional Responsibilities," *Yale J. Health Pol'y, L., & Ethics* 6 (2006): 269, 289; Nancy Berlinger, "Martin Luther at the Bedside," *Hastings Ctr. Rep.* 37, no. 2 (2007): 49.

Chapter 4

1. Pope John Paul II observed that "freedom of conscience does not confer a right to indiscriminate recourse to conscientious objection. When an asserted freedom turns into license or becomes an excuse for limiting the rights of others, the State is obliged to protect, also by legal means, the inalienable rights of its citizens against such abuses." Pope John Paul II, "If You Want Peace, Respect the Conscience of Every Person," January 1, 1993, www.vatican.va. While the Pope was referring to the dangers of political fundamentalism in denying human rights, his comments are quite apt to the situation facing patients when the physician's refusal truly would be a license to deny patient access to legal medical services.

2. Robert Vischer, "The Pharmacist Wars," *American Enterprise Online*, February 14, 2006, www.taemag.com; Carl H. Coleman, "Conceiving Harm: Disability Discrimination in Assisted Reproductive Technologies," *UCLA L. Rev.* 50 (2002): 17, 67.

3. See Kent Greenawalt, "Objections in Conscience to Medical Procedures: Does Religion Make a Difference?," *U. Ill. L. Rev.* (2006): 799, 823; Elizabeth Fenton and Loren Lomasky, "Dispensing with Liberty: Conscientious Refusal and the 'Morning-After Pill,'" *J. Med. & Phil.* 30, no. 6 (2005): 579, 588; Bernard M. Dickens and Rebecca J. Cook, "The Scope and Limits of Conscientious Objection," *Int'l J. Gynecology & Obstetrics* 71, no. 1 (2000): 71, 79; ACLU Reproductive Freedom Project, "Religious Refusals and Reproductive Rights," 2002, 8, www.aclu.org.

4. ACLU, "Religious Refusals," 11.

5. Catholics for Free Choice, "Position Paper on Individual and Institutional Refusal Clauses," April 2005, www.catholicsforchoice.org. See also R. Alta Charo, "The Celestial Fire of Conscience—Refusing to Deliver Medical Care," *New Eng. J. Med.* 352, no. 24 (2005): 2471; Rebecca Dresser, "Professionals, Conformity, and Conscience," *Hastings Ctr. Rep.* 35, no. 6 (2005): 9, 10.

6. These issues are discussed in greater detail in chapter 8.

7. Ill. Admin. Code tit. 68, § 1330.91 (2005).

8. See *Vandersand v. Wal-Mart Stores, Inc.*, in the U.S. District Court for the Central District of Illinois at the time this book went to press, involving a challenge based on this catch-22.

9. See Thom Wilder, "Pharmacy Regulation: Illinois Reaches Settlements with Pharmacists Over Dispensing of Plan B," *BNA Pharmaceutical Law & Industry Rep.* 5, no. 41 (October 19, 2007); Judy Peres, "'Morning-After' Pill Deal Reached," *Chicago Tribune*, October 11, 2007. The state has until March 2008

to make the regulatory changes, but because regulatory changes are involved, the settlement must be reviewed by a state panel before it can take effect.

10. The regulatory language provides: "If the contraceptive, or a suitable alternative, is not in stock, the pharmacy must obtain the contraceptive under the pharmacy's standard procedures for ordering contraceptive drugs not in stock, including the procedures of any entity that is affiliated with, owns, or franchises the pharmacy." Ill. Admin. Code tit. 68, § 1330.91(j)(1) (2005). The regulation requires pharmacies to post a visible notice making patients aware of their rights under the rule, but this notice is stated in conditional language: "If this pharmacy dispenses prescription contraceptives, then you have the following rights under Illinois law . . . ." Ill. Admin. Code tit. 68, § 1330.91(k)(2).

11. Wash. Admin. Code § 246-869-010 (2007).

12. *See* Washington Board of Pharmacy, "Rule Summary and Compliance Document," https://fortress.wa.gov/doh.

13. Wash. Admin. Code § 246-863-095 (2007).

14. W.D.Wa., Case No. C07-5374RBL, Nov. 8, 2007, www.alliancedefensefund .org/UserDocs/StormansPIRuling.pdf.

15. Cal. Bus. & Prof. Code §§ 733(b)(3) (West 2005).

16. National Women's Law Center, "Pharmacy Refusals: State Laws, Regulations, and Policies," 2006, 2, www.nwlc.org.

17. S. 809, 109th Cong. (2005); H.R. 1652, 109th Cong. (2005).

18. ACLU, "Religious Refusals," 23.

19. 517 A.2d 886, 889–890 (Super. Ct. N.J. 1986).

20. 517 A.2d at 893.

21. 529 A.2d 434 (N.J. 1987).

22. 697 F. Supp. 580, 591 (D.R.I. 1988).

23. 183 A.D.2d 10 (N.Y. App. Div. 1992).

24. See, for example, Michael A. Rie, "Defining the Limits of Institutional Moral Agency in Health Care: A Response to Kevin Wildes," *J. Med. Phil.* 16, no. 2 (1991): 221.

25. Edmund D. Pellegrino, "The Physician's Conscience, Conscience Clauses, and Religious Belief: A Catholic Perspective," *Fordham Urb. L.J.* 30 (2002): 221, 235–236.

26. Lynn D. Wardle, "Protecting the Rights of Conscience of Health Care Providers," *J. Legal Med.* 14 (1993): 177, 185.

27. Fifteen of these states limit the exemption to private health-care institutions and one state allows only religious health-care entities to refuse to provide such care. See Guttmacher Institute, "State Policies in Brief: Refusing to Provide Health Services," 2006, 1, www.guttmacher.org.

28. Six of these states limit the exemption to private entities. Guttmacher Institute, "State Policies in Brief."

29. Four of these states limit the exemption to private entities. Guttmacher Institute, "State Policies in Brief."

30. See the further discussion of this issue in chapter 9.

31. Norman Daniels, *Just Health: A Population View* (Cambridge: Cambridge University Press, 2007), 257.

32. R. Alta Charo, "Health Care Provider Refusals to Treat, Prescribe, Refer, or Inform: Professionalism and Conscience" (American Constitution Society, February 2007), 16–17, www.americanconstitutionsociety.org.

33. David Boddiger, "AMA Hopes Streamlined Agenda Will Boost Membership," *Lancet* 366, no. 9490 (2005): 971, 971.

34. See Guttmacher Institute, "State Policies in Brief: Restricting Insurance Coverage of Abortion," 2006, www.guttmacher.org (noting that several states restrict insurance coverage of abortion); Adam Sonfield, Rachel Benson Gold, Jennifer J. Frost, and Jacqueline E. Darroch, "U.S. Insurance Coverage of Contraceptives and the Impact of Contraceptive Coverage Mandates," *Perspectives on Sexual & Reproductive Health* 36 (2004): 72, 72 (indicating that while most plans do cover contraceptives, as many as 10 percent have some limitations); Tarun Jain, Bernard L. Harlow, and Mark D. Hornstein, "Insurance Coverage and Outcomes of in Vitro Fertilization," *New Eng. J. Med.* 347, no. 9 (2002): 661, 661 (noting that although most insurance companies in the United States do not cover in vitro fertilization, a few states, including Illinois, Massachusetts, Rhode Island, New Jersey, Arkansas, Hawaii, Maryland, Ohio, and West Virginia, mandate full or partial insurance coverage of this service).

35. Connie Marshner, "The Health Insurance Exchange: Enabling Freedom of Conscience in Health Care," March 1, 2007, www.heritage.org.

36. Further, the government's "conscience" would itself be defined by the very services for which licensing boards have an obligation to satisfy patient demand. Since they need not provide physicians willing to offer other services, the board's "conscience," to the extent it exists at all, is safe.

37. Robert E. Goodin, *Protecting the Vulnerable: A Reanalysis of Our Social Responsibilities* (Chicago: University of Chicago Press, 1985), 136.

38. Given the problems associated with placing the burden of accommodating both patients and professionals on employers, state pharmacy boards and other professional licensing agencies, such as nursing boards, could take on precisely the role recommended throughout this book for physician licensing boards. Of course, this role is quite complex and so the employer approach may still be preferable where it is available, even when it does not technically reflect the profession itself fulfilling its collective obligations. In the context of physician refusals, on the other hand, there is no easier solution, and in that case, it is an added validation of a complex proposal that the chosen institutional mediator represents the profession as a whole, such that the profession itself can be seen as satisfying its responsibility to those who are uniquely vulnerable to it.

39. Bernard M. Dickens, "Ethical Misconduct by Abuse of Conscientious Objection Laws," *Med. & Law* 25, no. 3 (2006): 513, 513 (mentioning this compromise exclusively in the article's abstract).

40. Dickens, "Ethical Misconduct," 515. Another vague suggestion of an institutional compromise has been offered by Patricia L. Selby, who in response to a proposed expansion of Michigan's conscience clause, stated that "the oversight responsibility of the Michigan Department of Community Health should be expressly expanded to monitor the impact of the law for potential loss of care—by specific types of health care as well as by region—requiring it to address any loss of services." Selby, "On Whose Conscience? Patient Rights Disappear under Broad Protective Measures for Conscientious Objectors in Health Care," *U. Det. Mercy L. Rev.* 83 (2006): 507, 520. Unfortunately, Selby does not indicate how such monitoring should occur, what standards of measurement should be used, or how the Department could or should address any loss of services.

41. Dickens and Cook, "Scope and Limits of Conscientious Objection," 76.

Chapter 5

1. Robert M. Veatch, *The Patient-Physician Relation: The Patient as Partner, Part 2* (Bloomington: Indiana University Press, 1991), 151–153.

2. David C. Goodman, Elliott S. Fisher, Thomas A. Bubolz, Jack E. Mohr, James F. Posage, and John E. Wennberg, "Benchmarking the U.S. Physician Workforce: An Alternative to Needs-Based or Demand-Based Planning," *J. Am. Med. Ass'n* 276, no. 22 (1996): 1811.

3. This issue is discussed in greater detail in chapter 7.

4. Linda O'Brien-Pallas, Andrea Baumann, Gail Donner, Gail Tomblin Murphy, Jacquelyn Lochhaas-Gerlach, and Marcia Luba, "Forecasting Models for Human Resources in Health Care," *J. Advanced Nursing* 33, no. 1 (2001): 120.

5. For example, the Supreme Court held in *Harris v. McRae*, 448 U.S. 297 (1980), that while a woman does have a right to freedom of choice with regard to abortion, a state's decision not to finance abortion services is not an undue burden on that freedom because the woman has no "constitutional entitlement to the financial resources to avail herself of the full range of protected choices."

6. For a discussion of the Ashley treatment provided by Ashley's parents, see http://ashleytreatment.spaces.live.com, and for criticisms see Nancy Gibbs, "Pillow Angel Ethics, Part 2," *Time* magazine, January 9, 2007, 56–57.

7. Most providers of preimplantation genetic diagnosis find these disability-producing requests unacceptable and will not satisfy them, although they might be willing to refer the patient elsewhere, and 3 percent of fertility clinics report providing this service. See Darshak M. Sanghavi, "Wanting Babies Like Themselves, Some Parents Choose Genetic Defects," *New York Times*, December 6, 2006.

8. See "Growth Hormones for Kids?," *CBS News*, June 12, 2003, www
.cbsnews.com.

9. Vanessa Renée Casavant, "Catman's Transformation Raises Concerns over
Extreme Surgery," *Seattle Times*, August 16, 2005, http://seattletimes.nwsource
.com.

10. See John Springer, "Woman Defends Decision to Give Birth at 60," *Today
Show*, May 24, 2007, www.todayshow.com.

11. Only a few months earlier, the *Today Show* featured a story about another
sixty-year-old woman who gave birth, and in December 2006, a sixty-seven-year-
old Spanish woman gave birth to twins after undergoing in vitro fertilization in
the United States. "67-Year-Old Spaniard Gives Birth to Twins," www.cnn.com,
December 30, 2006.

12. That men can father children into their old age or that some people who are
likely unfit to care for children are nonetheless able to have them does not necessarily make such behavior right, and these examples should probably not be used
to set the standard. Instead, they represent situations that society has determined
would be too intrusive on individual liberty to prohibit. However, where active
assistance is involved, society has generally been willing to impose heightened
restrictions, as exemplified by the adoption system's in-depth screening of potential parents. Whether those heightened restrictions are appropriate, and if so,
how far they should go, are issues at the heart of this chapter.

13. Arthur Caplan, "Media's Cooing over Sextuplets Is a Disservice," *MSNBC*,
June 19, 2007, www.msnbc.com; Lorie A. Parch, "2-for-1 Is No Baby Bargain,
Doctors Say," *MSNBC*, June 19, 2007, www.msnbc.com.

14. Gregory E. Pence, "The McCaughey Septuplets: God's Will or Human
Choice?," in Helga Khuse and Peter Singer, eds., *Bioethics: An Anthology*, 2nd
ed. (Oxford: Blackwell, 2006), 88. For a fascinating and emotional discussion of
selective reduction, see Liza Mundy, "Too Much to Carry," *Washington Post*,
May 20, 2007, W14.

15. The American Society for Reproductive Medicine has issued guidelines for
the fertility industry based on patient age: women under thirty-five should never
be given more than two embryos during a single implantation procedure, absent
extraordinary circumstances; women age thirty-five to thirty-seven should receive
no more than three embryos; women age thirty-eight to forty should receive no
more than four; and women over forty should receive no more than five. Practice
Committee of the Society for Assisted Reproductive Technology and the Practice Committee of the American Society for Reproductive Medicine, "Guidelines
on Number of Embryos Transferred," *Fertility & Sterility* 86 (supp. 4) (2006):
S51. Physicians refusing to implant more embryos than described in these guidelines are therefore relying on professional standards, but note that while these
professional standards are based at least partly on medical expertise, they also include value judgments with which patients could reasonably disagree. Thus, they

nonetheless represent a type of conscientious refusal that will be discussed in more detail later.

16. Parch, "2-for-1."

17. Bernard M. Dickens, "Ethical Misconduct by Abuse of Conscientious Objection Laws," *Med. & Law* 25, no. 3 (2006): 513, 515.

18. Veatch, *The Patient-Physician Relation*, 152.

19. See, for example, Björn J. Oddens, "Women's Satisfaction with Birth Control: A Population Survey of Physical and Psychological Effects of Oral Contraceptives, Intrauterine Devices, Condoms, Natural Family Planning, and Sterilization among 1466 Women," *Contraception* 59, no. 5 (1999): 277.

20. Richard J. Fehring, Mary Schneider, Kathleen Raviele, and Mary Lee Barron, "Efficacy of Cervical Mucus Observations Plus Electronic Hormonal Fertility Monitoring as a Method of Natural Family Planning," *J. Obstetric, Gynecologic, & Neonatal Nursing* 36, no. 2 (2007): 152.

21. American Medical Association, "Code of Medical Ethics," § 10.05.

22. Alexander M. Capron, "In Re Helga Wanglie," *Hastings Ctr. Rep.* 21, no. 5 (1991): 26. *See also* Robert M. Veatch and Carol Mason Spicer, "Medically Futile Care: The Role of the Physician in Setting Limits," *Am. J. L. & Med.* 18, no. 1–2 (1992): 15, 23; Tom L. Beauchamp and James Childress, *The Principles of Biomedical Ethics*, 5th ed. (Oxford: Oxford University Press, 2001), 192.

23. Sylvia Moreno, "Case Puts Texas Futile-Treatment Law under a Microscope," *Washington Post*, April 11, 2007, A3. The states are Texas and Virginia. See, for example, Texas Advance Directives Act, §166.046 (1999).

24. John Stuart Mill, "On Liberty," in *The Basic Writings of John Stuart Mill* (New York: Modern Library, 2002), 77–96. Mill states that when "a person's conduct affects the interests of no persons besides himself, or needs not affect them unless they like ... there should be perfect freedom, legal and social, to do the action and stand the consequences." However, he limits this principle to persons of "full age" who bear the "ordinary amount" of understanding. Mill, "On Liberty," 78.

25. Mill, "On Liberty," 100.

26. But note that it cannot be the case that patients requesting these services are considered irrational simply because rational patients would not make such requests; of course, that sort of reasoning is entirely tautological. As we saw earlier, it is possible that even if a patient was fully aware of the personal risks of female circumcision, she could logically choose to accept those risks for herself, arguing that without the procedure, her marital prospects and social status would dwindle, even if they absolutely should not in an ideal world. Similarly, a patient suffering horribly through a chronic illness might rationally choose to die.

27. Council of the Society of Obstetricians and Gynecologists of Canada, "Gender Selection," www.csdms.com.

28. See Mill's discussion of "social rights" in "On Liberty," 92–93.

29. Others have claimed, albeit on shaky empirical grounds, that women themselves are harmed both physically and emotionally by their choice to terminate a pregnancy. However, so long as women are informed of these risks, this reasoning could not be used to exclude abortion under the harm principle. See Robin Toner, "Abortion Foes See Validation for New Tactic," *New York Times*, May 22, 2007.

30. See Norman Daniels, "Chapter 2: What Is the Special Moral Importance of Health?," *Just Health: A Population View* (Cambridge: Cambridge University Press, 2007).

31. R. Alta Charo, "The Celestial Fire of Conscience—Refusing to Deliver Medical Care," *New Eng. J. Med.* 352, no. 24 (2005): 2471, 2473.

32. Daniels, *Just Health*, 41.

33. See, for example, Sean Murphy, "Freedom of Conscience and the Needs of the Patient," presentation at 2001 Obstetrics and Gynecology Conference: New Developments–New Boundaries, Banff, Alberta, 2 n. 18, www.consciencelaws.org.

34. See, for example, John F. Kilner, "Healthcare Resources, Allocation of; Macroallocation," in Stephen G. Post, ed., *The Encyclopedia of Bioethics*, 3rd ed. (New York: Macmillan Reference, 2004), 1101.

35. Einer Elhauge, "Allocating Health Care Morally," *Cal. L. Rev.* 82 (1994): 1449, 1467–1468.

36. Carl Elliott describes this mental disorder in "Amputees by Choice," chapter 9 of his book *Better Than Well: American Medicine Meets the American Dream* (New York: Norton, 2003).

37. This is precisely the argument made by Deirdre McQuade, planning director for the U.S. Conference of Catholic Bishops' Secretariat for Pro-Life Activities, with regard to the prescription and sale of Plan B: "Pregnancy is not a disease.... There is no absolute duty to dispense a non-therapeutic drug, but there is a basic civil right of conscience." David Crary, "Sales Soar for Morning-After Pill," Associated Press, August 22, 2007.

38. Charu A. Chandrasekhar, "RX for Drugstore Discrimination: Challenging Pharmacy Refusals to Dispense Prescription Contraceptives under State Public Accommodation Laws," *Alb. L. Rev.* 70 (2006): 55, 63–65; Claire A. Smearman, "Drawing the Line: The Legal, Ethical, and Public Policy Implications of Refusal Clauses for Pharmacists," *Ariz. L. Rev.* 48 (2006): 469, 473.

39. Charo, "The Celestial Fire of Conscience," 2473.

40. Mark R. Wicclair, "Conscientious Objection in Medicine," *Bioethics* 14, no. 3 (2000): 205, 206.

41. For an excellent discussion of these barriers in the context of passing the Civil Rights Act of 1964, see William N. Eskridge Jr., Philip P. Frickey, and Elizabeth

Garrett, eds., *Cases and Materials on Legislation: Statutes and the Creation of Public Policy*, 3rd ed. (St. Paul, MN: West Publishing Co. 2001), 4–23.

42. Teresa Stanton Collett, "The Common Good and the Duty to Represent: Must the Last Lawyer in Town Take Any Case?," *S. Tex. L. Rev.* 40 (1999): 137, 167.

43. John F. Manning, "Lawmaking Made Easy," *Green Bag 2d* 10 (2007): 191, 193 (discussing *J.W. Hampton and Co. v. U.S.*, 276 U.S. 394, 409 (1928), in which the court noted that forcing Congress itself to fix every rate to be exacted by interstate carriers for passenger and merchandise traffic would make it effectively impossible for Congress to exercise its rate-making power).

44. But note that not everyone would agree that the efficiency gained by delegating power to licensing boards to further restrict which medical services are available to patients is appropriate. They would argue that the fact that legislatures have difficulty rendering medical services illegal is actually for the best, since it preserves the maximum freedom for patients. Proponents of this "less is more" approach to lawmaking would simply advocate the legality/scientific expertise standard as the ideal limit on licensing board obligations. See Manning, "Lawmaking Made Easy," 198–204.

45. While some argue that the medical profession has "moral and not merely empirical authority in moral-medical questions," most commentators rightly disagree. See, for example, Lynn A. Jansen, "No Safe Harbor: The Principle of Complicity and the Practice of Voluntary Stopping of Eating and Drinking," *J. Med. & Phil.* 29, no. 1 (2004): 61, 71; M. T. Harvey, "What Does a 'Right' to Physician-Assisted Suicide (PAS) Legally Entail?," *Theoretical Med. & Bioethics* 23, no. 4–5 (2002): 271, 276; J. David Bleich, "The Physician as a Conscientious Objector," *Fordham Urb. L.J.* 30 (2002): 245, 257; FrançoiseBaylis, "Expert Testimony by Persons Trained in Ethical Reasoning: The Case of Andrew Sawatzky," *J.L. Med & Ethics* 28 (2000): 224, 229; Edmund D. Pellegrino, "Patient and Physician Autonomy: Conflicting Rights and Obligations in the Physician-Patient Relationship," *J. Contemp. Health L. & Pol'y* 10 (1993): 47, 68; Veatch and Spicer, "Medically Futile Care," 36.

46. Julian Savulescu, "Conscientious Objection in Medicine," *Brit. Med. J.* 332 (2006): 294, 294.

47. Talcott Parsons, *The Social System* (London: Routledge, 1951), 435.

48. John F. Crosby, "There Is No Moral Authority in Medicine: Response to Cowdin and Tuohey," *Christian Bioethics* 4, no. 1 (1998): 63, 63.

49. Robert M. Veatch, "Doctor Does Not Know Best: Why in the New Century Physicians Must Stop Trying to Benefit Patients," *J. Med. & Phil.* 25, no. 6 (2000): 701, 705.

50. For arguments on both sides of the debate regarding whether moral expertise can exist at all, see Bruce D. Weinstein, "The Possibility of Ethical Expertise," *Theoretical Med. & Bioethics* 15, no. 1 (1994): 61. See also Ronald Suter, *Are You Moral?* (Lanham, MD: University Press of America, 1984), 56; C. D. Broad, *Ethics and the History of Philosophy* (London: Routledge, 1952), 244;

Christopher Crowley, "A New Rejection of Moral Expertise," *Med., Health Care & Phil.* 8, no. 3 (2005): 273, 276; Scott D. Yoder, "The Nature of Ethical Expertise," *Hastings Ctr. Rep.* 28, no. 6 (1998): 11. However, even if substantive moral expertise is nonexistent, a person could be an expert in the sense of being more equipped to provide strong justifications for normative conclusions. See, for example, Yoder, "Ethical Expertise"; Weinstein, "Ethical Expertise," 68; Peter Singer, "Moral Expertise," *Analysis* 32 (1972): 115, 117.

51. John K. Davis, "Conscientious Refusal and a Doctor's Right to Quit," *J. Med. & Phil.* 29, no. 1 (2004): 75, 85–86.

52. 497 N.E.2d 626, 639 (Mass. 1986).

53. Pellegrino, "Patient and Physician Autonomy," 53.

54. Mark Strasser, "The Futility of Futility? On Life, Death, and Reasoned Public Policy," *Md. L. Rev.* 57 (1998): 505, 547.

55. Alexander M. Capron, "Professionalism and Professional Ethics," in Robert B. Baker, Arthur L. Caplan, Linda L. Emanuel, and Stephen R. Latham, eds., *The American Medical Ethics Revolution* (Baltimore: Johns Hopkins University Press, 1999), 185.

56. Robert M. Veatch, "Who Should Control the Scope and Nature of Medical Ethics?," in Robert B. Baker, Arthur L. Caplan, Linda L. Emanuel, and Stephen R. Latham, eds., *The American Medical Ethics Revolution* (Baltimore: Johns Hopkins University Press, 1999), 162.

57. Martha Swartz, "Conscience Clauses or Unconscionable Clauses: Personal Beliefs versus Professional Responsibilities," *Yale J. Health Pol'y, L., & Ethics* 6 (2006): 269, 313–314.

58. Physicians for Life, "U.S. Physicians' Right of Conscience Amendment Under Attack—Again," September 2005, www.physiciansforlife.org.

59. Daniels, *Just Health*, 113–155.

60. However, it is possible that this approach could allow some services to be inappropriately excluded given that the standard of care is determined by the usual practice of physicians themselves.

61. Daniels, *Just Health*, 113–115.

62. 5 U.S.C. § 500 et seq.

63. For a concise discussion of federal agency rulemaking procedure, see Richard J. Pierce Jr., Sidney A. Shapiro, and Paul R. Verkuil, *Administrative Law and Process*, 3rd ed. (New York: Foundation Press, 1999), 320–331.

64. Daniels, *Just Health*, 122.

65. Daniels, *Just Health*, 122–123.

66. Daniels, *Just Health*, 143.

67. David M. Mirvis, "Physicians' Autonomy—The Relation between Public and Professional Expectations," *New. Eng. J. Med.* 328, no. 18 (1993): 1346, 1346.

Chapter 6

1. Cal. Bus. & Prof. Code § 733 (b)(3) (2007).

2. Judith F. Daar, "A Clash at the Bedside: Patient Autonomy v. a Physician's Professional Conscience," *Hastings L.J.* 44 (1993): 1241, 1287.

3. Texas Advance Directive Act, §§ 166.046 and 166.052.

4. Texas Advance Directive Act, § 166.053.

5. *Walker v. Pierce*, 560 F.2d 609 (4th Cir. 1977).

6. Mark R. Wicclair, "Conscientious Objection in Medicine," *Bioethics* 14, no. 3 (2000): 205, 221; Christopher Meyers and Robert D. Woods, "An Obligation to Provide Abortion Services: What Happens When Physicians Refuse?," *J. Med. Ethics* 22, no. 2 (1996): 115, 119; Christopher L. Eisgruber and Lawrence G. Sager, "The Vulnerability of Conscience: The Constitutional Basis for Protecting Religious Conduct," *U. Chi. L. Rev.* 61 (1994): 1245, 1291.

7. Kent Greenawalt, "Objections in Conscience to Medical Procedures: Does Religion Make a Difference?," *U. Ill. L. Rev.* (2006): 799, 800.

8. Eisgruber and Sager, "The Vulnerability of Conscience," 1263.

9. Mark R. Wicclair, "Reasons and Health Care Professionals' Claims of Conscience," *Am. J. Bioethics* 7, no. 6 (2007): 21.

10. Piers Benn, "The Role of Conscience in Medical Ethics," in Nafiska Athanssoulis, ed., *Philosophical Reflections on Medical Ethics* (New York: Palgrave Macmillan, 2005), 177.

11. *Cutter v. Wilkinson*, 544 U.S. 709, 718–719 (2005) (rejecting the argument that privileging religion over other deeply held beliefs is unconstitutional).

12. 50 U.S.C. § 456(j).

13. *U.S. v. Seeger*, 380 U.S. 164, 185 (1965). The Court explained that the test of acceptable conscientious objection to the military draft "might be stated in these words: A sincere and meaningful belief which occupies in the life of its possessor a place parallel to that filled by the God of those admittedly qualifying for the exemption comes within the statutory definition." *Seeger*, 380 U.S. at 176.

14. Note that some commentators recognize that allocation of scarce medical resources may be a reason to treat some patients differently from others when they differ in relevant respects, but they refuse to accept that such decisions can be properly made by individual physicians for three major reasons: (1) unless allocation decisions are made on a broader social level, similarly situated patients could end up being treated differently solely as a result of their choice of physician; (2) individual physicians have a duty of loyalty to their patients and must act in their best interests without regard to competing claims from others; and (3) individual doctors lack any special expertise or knowledge to make these value-based choices. See Jack Coulehan, Peter C. Williams, S. Van McCrary, and Catherine Belling, "The Best Lack All Conviction: Biomedical Ethics, Professionalism, and Social Responsibility," *Cambridge Q. Healthcare Ethics* 12, no. 1

(2003): 21, 24; David Orentlicher, "Trends in Health Care Decisionmaking: The Limits of Legislation," *Md. L. Rev.* 53 (1994): 1255, 1297–1298; Robert M. Veatch and Carol Mason Spicer, "Medically Futile Care: The Role of the Physician in Setting Limits," *Am. J. L. & Med.* 18, no. 1–2 (1992): 15, 28–29; Marion Danis and L. R. Churchill, "Autonomy and the Common Weal," *Hastings Ctr. Rep.* 21, no. 1 (1991): 25. See also Julian Savulescu, "Conscientious Objection in Medicine," *Brit. Med. J.* 332 (2006): 294, 295 (arguing that it would be inappropriate for an individual physician to refuse to treat people over the age of seventy on the grounds that they have already had "fair innings" unless society had also determined that such patients are not entitled to treatment).

15. Elizabeth Fenton and Loren Lomasky, "Dispensing with Liberty: Conscientious Refusal and the 'Morning-After Pill,'" *J. Med. & Phil.* 30, no. 6 (2005): 579, 584.

16. Jacob M. Appel, "May Doctors Refuse Infertility Treatments to Gay Patients?," *Hastings Ctr. Rep.* 36, no. 4 (2006): 20, 21.

17. Robert K. Vischer, "Conscience in Context: Pharmacist Rights and the Eroding Moral Marketplace," *Stan. L. & Pol'y Rev.* 17 (2006): 83, 89. But note that it is more than mere embarrassment when a person who has already been through an incredibly harrowing personal experience is then refused services, such as the rape victim who is refused emergency contraception in the hospital. This is the sort of situation where matching is impossible and special obligations may come into play. See the discussion of this issue in chapter 9.

18. William W. Bassett, "Private Religious Hospitals: Limitations Upon Autonomous Moral Choices in Reproductive Medicine," *J. Contemp. Health L. & Pol'y* 17 (2001): 455, 535.

19. R. Alta Charo, "Health Care Provider Refusals to Treat, Prescribe, Refer, or Inform: Professionalism and Conscience" (American Constitution Society, February 2007), 12, www.americanconstitutionsociety.org.

20. Charu A. Chandrasekhar, "RX for Drugstore Discrimination: Challenging Pharmacy Refusals to Dispense Prescription Contraceptives Under State Public Accommodation Laws," *Alb. L. Rev.* 70 (2006): 55, 63–65. See also Dana E. Blackman, "Refusal to Dispense Emergency Contraception in Washington State: An Act of Conscience or Unlawful Sex Discrimination?," *Mich J. Gender & L.* 14 (2007): 59.

21. Julian Savulescu, "Conscientious Objection in Medicine," 297.

22. Similarly, patients should not be permitted to demand particular types of physicians on invidiously discriminatory grounds; such demands are the sort that can be appropriately ignored by licensing boards. See the discussion of this issue in the second part of chapter 3.

23. For example, the Civil Rights Act of 1964 prohibits physicians and hospitals receiving federal funding, including Medicare and Medicaid, from discriminating against patients and potential patients on the basis of race, color, religion, or na-

tional origin. Appel notes that many states have expanded these protections to cover gender and sexual orientation as well. Appel, "May Doctors Refuse," 20.

24. Charo, "Health Care Provider Refusals to Treat, Prescribe, Refer, or Inform," 11.

25. American Medical Association, "Code of Medical Ethics," § 10.05, based on Opinion § 9.12 ("physicians who offer their services to the public may not decline to accept patients because of race, color, religion, national origin, sexual orientation, or any other basis that would constitute invidious discrimination"). See also Robert M. Sade, American Medical Association, "Report of the Council on Ethical and Judicial Affairs 6-A-07: Physician Objection to Treatment and Individual Patient Discrimination," referred to Committee on Amendments to Constitution and Bylaws (2007), http://www.ama-assn.org/ama1/pub/upload/mm/369/ceja_6a07.pdf.

26. World Medical Association, "International Code of Medical Ethics and Declaration of Geneva," www.wma.net.

27. North Coast Women's Medical Care Group, Health Law Amicus Brief, 43–44, www.lambdalegal.org.

28. Miss. Code Ann. § 41-107-5(1) (2005).

29. *Gillette v. U.S.*, 401 U.S. 437 (1971).

30. See, for example, L. Ebony Boulware, Lisa A. Cooper, Lloyd E. Ratner, Thomas A. LaVeist, and Neil R. Powe, "Race and Trust in the Health Care System," *Public Health Rep.* 118, no. 4 (2003): 358.

31. Appel, "May Doctors Refuse," 21.

32. Thomas May, "Rights of Conscience in Health Care," *Soc. Theory & Practice* 27, no. 1 (2001): 111, 113.

33. Eisgruber and Sager, "The Vulnerability of Conscience," 1269.

34. While the AMA ultimately rejected this approach, some doctors have suggested refusing treatment to malpractice attorneys and their families barring an emergency. Fred Charatan, "US Doctors Debate Refusing Treatment to Malpractice Lawyers," *Brit. Med. J.* 328 (2004): 1518.

35. Similarly, Samuel Gorovitz argues that a physician's feelings of repugnance, contempt, dislike, or loathing toward a patient should not influence the physician's behavior toward or willingness to treat a homeless person whom the doctor believes should be working, a patient who publicly criticizes the physician's religion, a pickpocket, or a single or lesbian woman seeking to become pregnant. Samuel Gorovitz, *Doctors' Dilemmas: Moral Conflict and Medical Care* (Oxford: Oxford University Press, 1982), 99–105.

36. "Tattooed Mom Criticizes Doc's Refusal to Help Her Sick Child," *San Diego Union-Tribune*, March 15, 2007.

37. A refusal to prescribe Viagra to homosexual patients experiencing erectile dysfunction would also be impermissible.

38. "There is no persuasive evidence that children raised by single parents or by gays and lesbians are harmed or disadvantaged by that fact alone." American Society for Reproductive Medicine, Ethics Committee, "Access to Fertility Treatment by Gays, Lesbians, and Unmarried Persons," *Fertility & Sterility* 86, no. 5 (2006): 1333.

39. Note, however, that if the physician refused to implant embryos in all women beyond normal reproductive age or in all women who are not clinically infertile, that might be acceptable even though the physician does not necessarily object to IVF as a practice, since these could be relevant characteristics that are at the very least separate from moral castigation of the patient as a person.

40. Andrea D. Gurmankin, Arthur L. Caplan, and Andrea M. Braverman, "Screening Practices and Beliefs of Assisted Reproductive Technology Programs," *Fertility & Sterility* 83, no. 1 (2005): 61.

41. American Society for Reproductive Medicine, Ethics Committee, "Access to Fertility Treatment," 1335.

42. The case has been accepted for review by the California State Supreme Court as *North Coast Women's Care Medical Group, Inc. v. Superior Court*, but was previously referred to as *Benitez v. North Coast Women's Care Medical Group*.

43. Bob Egelko, "State High Court to Hear Lesbian's Case; Doctors Denied Her Infertility Treatment on Religious Grounds," *San Francisco Chron.*, June 15, 2006, www.sfgate.com.

44. *Benitez v. N. Coast Women's Care Med. Group, Inc.* (Cal. App. 4th Dist. 2003).

45. Lambda Legal, "Benitez v. North Coast Women's Care Medical Group (Q & A)," www.lambdalegal.org.

46. Cal. Health & Safety Code § 123420 (West 2006).

47. Egelko, "Lesbian's Case."

48. Kenneth Ofgang, "State Supreme Court Agrees to Hear Lesbian Insemination Case," *Metropolitan News-Enterprise*, June 15, 2006, www.metnews.com.

49. *N. Coast Women's Care Med. Group, Inc. v. Superior Ct.* (Cal. App. 4th Dist. 2005).

50. Note, however, that because black, female, and gay patients would likely avoid a doctor known to be racist, sexist, or homophobic, or to exhibit other discriminatory tendencies, even if that doctor could not necessarily reject them, he may nevertheless be able to avoid situations that he finds objectionable. What is important is that should a patient want that doctor's services, the doctor will be without protection should he refuse.

51. Robert Card has also suggested that conscientious refusers in the health-care setting should have to give reasons and their reasons should be open to evaluation in a fashion similar to that used for objections to military service. Card, "Conscientious Objection and Emergency Contraception," *Am. J. Bioethics* 7, no. 6 (2007): 8, 13. Similarly, Meyers and Woods explain that because profes-

sionals are not "free agents" and owe duties to the public and to patients, they cannot just say no, but instead have some burden of proof to show why they should be exempted from those duties. They go on to argue that this burden can be satisfied simply by offering a genuine conscientious objection, evaluated for sincerity by a review board similar to that employed by the draft board. Christopher Meyers and Robert D. Woods, "Conscientious Objection? Yes, But Make Sure It Is Genuine," *Am. J. Bioethics* 7, no. 6 (2007): 19, 20.

52. 380 U.S. 164, 184–185 (1965).

53. 398 U.S. 333, 339 (1970).

54. Fred C. Zacharias, "The Lawyer as Conscientious Objector," *Rutgers L. Rev.* 54 (2001): 191, 203–204. See also Martin Benjamin, "*Conscience*," in Stephen G. Post, ed., *The Encyclopedia of Bioethics*, 3rd ed. (New York: Macmillan Reference, 2004), 516.

55. For more information on conscientious objections to participation in the draft, see www.sss.gov, and Center on Conscience and War, "Conscientious Objectors and the Draft," 2001, www.centeronconscience.org.

56. Objectors to the draft must file the SSS Form 22, Claim Documentation Form—Conscientious Objector, which contains these questions.

57. Note that this is the sort of refusal that some would define as moral, but not rooted in conscience, given its weaker connection to personal integrity. Recall that for our purposes, however, all normative refusals are treated as conscientious refusals. See the discussion of this issue in the final section of chapter 1.

58. Greenawalt, "Objections," 820 n.88.

59. B. M. Dickens and R. J. Cook, "The Scope and Limits of Conscientious Objection," *Int'l J. Gynecology & Obstetrics* 71, no. 1 (2000): 71, 72.

60. Savulescu, "Conscientious Objection," 294.

61. May, "Rights of Conscience," 116 (emphasis added).

62. Jeffrey Blustein, "Infertility Treatments for Gay Parents?," letter to the editor, *Hastings Ctr. Rep.* 36, no. 5 (2006): 6.

63. Wicclair, "Reasons and Health Care Professionals' Claims of Conscience," 21.

Chapter 7

1. Elizabeth Fenton and Loren Lomasky, "Dispensing with Liberty: Conscientious Refusal and the 'Morning-After Pill,'" *J. Med. & Phil.* 30, no. 6 (2005): 579, 588–589.

2. Leslie Cannold, "Consequences for Patients of Health Care Professionals' Conscientious Actions: The Ban on Abortions in South Australia," *J. Med. Ethics* 20, no. 2 (1994): 80, 81.

3. 517 A.2d 886, 889 (Super. Ct. N.J. 1986).

4. 517 A.2d at 889.

5. Panel Discussion Hosted by the Pew Forum on Religion and Public Life, the American Constitution Society, and the Federalist Society, "Dr. No? The Debate on Conscience in Health Care," September 8, 2006, http://pewforum.org.

6. Fahui Wang and Wei Luo, "Assessing Spatial and Nonspatial Factors for Healthcare Access: Towards an Integrated Approach to Defining Health Professional Shortage Areas," *Health & Place* 11, no. 2 (2005): 131, 132–133.

7. R. Hewitt Pate, "Monopolization," *PLI/Corp.* 1541 (2006): 157, 159.

8. Justin B. Nelson, "Antitrust: Relevant Market Analysis of Health Care Providers," *Health L.* 9 (1997): 6, 6.

9. Nelson, "Antitrust," 6; U.S. Department of Justice and Federal Trade Commission, "Horizontal Merger Guidelines," § 1.21 (1992).

10. Pate, "Monopolization," 163.

11. *FTC v. Freeman Hospital*, 69 F.3d 260, 268 (8th Cir. 1995). See also "Monopolies, Restraints of Trade, and Unfair Trade Practices: What Constitutes the Geographic Market," 54 Am. Jur. 2d Monopolies, Restraints of Trade, Unfair Trade, § 57 (2006).

12. Thomas R. McCarthy and Scott J. Thomas, "Geographic Market Issues in Hospital Mergers," *The Health Care Mergers and Acquisitions Handbook* (American Bar Association, 2003), 2–4.

13. McCarthy and Thomas, "Geographic Market Issues," 12.

14. *Gordon v. Lewistown Hospital*, 272 F.Supp.2d 393, 429 (M.D. Pa. 2003); *Cogan v. Harford Memorial Hospital*, 843 F. Supp. 1013, 1019 (D.Md. 1994); *In re Evanston Northwestern Healthcare Corp.*, No. 9315 (FTC Initial Decision Oct. 20, 2005); and Pate, "Monopolization," 164.

15. McCarthy and Thomas, "Geographic Market Issues," 20.

16. McCarthy and Thomas, "Geographic Market Issues," 26.

17. Nelson, "Antitrust," 7. This is also known as the LOFI-LIFO test: "little out from inside, little in from outside."

18. The weaknesses of the tests for geographic markets as applied to the healthcare setting were described by Dr. Elzinga himself when he was called to testify by the FTC in a hospital merger antitrust challenge. Robert W. McCann, "Implications of the *Evanston Northwestern* Decision for Hospital Mergers and Health Care Antitrust Compliance," *Health Law Handbook* (Minneapolis: West Publishing, 2006), 735–736.

19. See, for example, *FTC v. Tenet Health Care Corp.*, 186 F.3d 1045, 1054 (8th Cir. 1999).

20. Stephanie Mueller and Susan Dudley, National Abortion Federation, "Access to Abortion," 2003, 1, www.prochoice.org.

21. "Abortion Ultrasound Bill Advances in S.C.; First-of-Its-Kind Law Would Require Women to See Ultrasound Before Having Abortion," *CBS News,* March

21, 2007, www.cbsnews.com. Of course, the procedure may also be performed in facilities not specifically devoted to its provision, such as hospitals or doctors' offices.

22. Martha Swartz, "Conscience Clauses or Unconscionable Clauses: Personal Beliefs versus Professional Responsibilities," *Yale J. Health Pol'y, L., & Ethics* 6 (2006): 269, 333.

23. The number of abortion providers has fallen 37 percent between 1982 and 2000, to only about 1,800 physicians, compared to 6,200 plastic surgeons, 9,700 dermatologists, and 10,600 gastroenterologists. Stephanie Simon, "Aspiring Abortion Doctors Drawn to Embattled Field," *LA Times*, May 22, 2007, www.latimes.com.

24. 351 U.S. 377 (1956).

25. *U.S. v. Eastman Kodak Co.*, 63 F.3d 95, 105 (2d Cir. 1995) ("At a high enough price, even poor substitutes look good to the consumer").

26. David C. Goodman, Elliott S. Fisher, Thomas A. Bubolz, Jack E. Mohr, James F. Posage, and John E. Wennberg, "Benchmarking the U.S. Physician Workforce: An Alternative to Needs-Based or Demand-Based Planning," *J. Am. Med. Ass'n* 276, no. 22 (1996): 1811.

27. For example, in *Gordon v. Lewistown Hospital*, 272 F.Supp.2d 393, 430 (M.D. Pa. 2003), the court noted that due to the severe impact of a cataract on a patient's life, the patient should be willing to travel at least some appreciable distance for treatment. Similarly, the court in *Blue Cross and Blue Shield of Wisconsin v. Marshfield Clinic*, 65 F.3d 1406, 1411 (7th Cir. 1995), stated that the market for primary care is generally local, but made the important qualification that people are willing to travel a long distance for an organ transplant.

28. Mark A. Hall, Mary Anne Bobinski, and David Orentlicher, *Health Care Law and Ethics*, 6th ed. (New York: Aspen Publishers, 2003), 1174. See also *U.S. v. Rockford Memorial Corp.*, 898 F.2d 1278 (7th Cir. 1990); Nelson, "Antitrust," 7.

29. John V. Jacobi, "Consumer-Directed Health Care and the Chronically Ill," *U. Mich. J.L. Reform* 38 (2005): 531, 537–538.

30. Ohio Admin. Code § 3701:12-20(E) (2004). See also Montana Code Ann. § 50-5-304(10) (2005).

31. Ohio Admin. Code § 3701:12-20(I).

32. Ohio Admin. Code § 3701:12-20(P).

33. 28 Pa. Code §§ 401.4(a)(13) and (14).

34. Fla. Stat. Ann. § 408.035 (West 2004).

35. Fla. Admin. Code Ann. r. 59C-1-030.

36. N.Y. Comp. Codes R. and Regs. tit. 10, §§ 709.1(b) and (c) (1997).

37. Ariz. Admin. Code § 4-23-611.

38. 5-19-1 R.I. Code R. § 13.1.3.

39. U.S. Health Resources and Services Administration, Bureau of Health Professions, "Shortage Designations," http://bhpr.hrsa.gov/shortage.

40. 42 C.F.R. § 5.2 (1993)—Appendix A, Part I—Geographic Areas, Section A.

41. 42 C.F.R. § 5.2 (1993)—Appendix A, Part I—Geographic Areas, Section B.1.a.

42. However, the regulation does provide mileage guidelines to be used when determining which distances correspond to thirty minutes travel time. 42 C.F.R. § 5.2 (1993)—Appendix A, Part I—Geographic Areas, Section B.1.b.

43. 42 C.F.R. § 5.2 (1993)—Appendix A, Part I—Geographic Areas, Section B.5.

44. 42 C.F.R. § 5.2 (1993)—Appendix A, Part I—Geographic Areas, Section B.4.

45. 42 C.F.R. § 5.2 (1993)—Appendix A, Part I—Geographic Areas, Section B.6.c.

46. For example, the HPSA regulations allow a greater burden on mental health patients than primary-care patients: forty minutes travel time to measure rational areas for delivery of mental health care versus only thirty for primary care. 42 C.F.R. § 5.2 (1993)—Appendix C, Part I—Geographic Areas.

47. Katherine R. Kruse, "Lawyers, Justice, and the Challenge of Moral Pluralism," *Minn. L. Rev.* 90 (2005): 389, 456.

48. John Wadleigh, "Clinical Case: Should I Stay or Should I Go? The Physician in a Time of Crisis," *AMA Virtual Mentor* 8 (2006): 208.

49. Christopher Meyers and Robert D. Woods, "An Obligation to Provide Abortion Services: What Happens When Physicians Refuse?," *J. Med. Ethics* 22, no. 2 (1996): 115, 120 n. 26.

50. Note that the gynecologist who is willing to provide abortions may truly be accepting more tangible risk and stress than her refusing colleagues. However, the physical risks associated with being a publicly known abortion provider are absolutely unacceptable and should not have to be borne by any physician. If no physician should be subject to this risk, the refuser is not inappropriately shirking his share of the burden; this is a problem beyond conscientious refusal that must be addressed by society as a whole.

51. Dennis Cauchon, "Medical Miscalculation Creates Doctor Shortage," *USA Today*, March 2, 2005, www.usatoday.com.

52. In fact, all specialties other than family medicine are more likely to be practiced in urban areas. Roger A. Rosenblatt and L. Gary Hart, "Physicians and Rural America," *West. J. Med.* 173, no. 5 (2000): 348, 348.

53. Howard K. Rabinowitz, James J. Diamond, Fred W. Markham, and Christina E. Hazelwood, "A Program to Increase the Number of Family Physicians in Rural and Underserved Areas," *J. Am. Med. Ass'n* 281, no. 3 (1999): 255, 255–256. As of 1999, one out of seventeen rural counties in the United States had no physician providing patient care, and out of those rural counties that did have a

physician, one-quarter had general and family practitioners exclusively and two-thirds had no pediatricians or obstetricians whatsoever. Laura Weiss Roberts, John Battaglia, Margaret Smithpeter, and Richard S. Epstein, "An Office on Main Street: Health Care Dilemmas in Small Communities," *Hastings Ctr. Rep.* 29, no. 4 (1999): 28, 34.

54. Steven A. Schroeder and Michael P. Beachler, "Physician Shortages in Rural America," *Lancet* 345, no. 8956 (1995): 1001, 1001.

55. "Influx of Doctors Overwhelms Texas Board," *Washington Post*, July 9, 2007.

56. U.S. Government Accountability Office "Health Professional Shortage Areas: Problems Remain with Primary Care Shortage Area Designation," October 2006, www.gao.gov.

57. Council on Graduate Medical Education, "10th Report: Physician Distribution and Health Care Challenges in Rural and Inner-City Areas," February 4, 1998, http://cogme.gov/10.pdf.

58. Council on Graduate Medical Education, "10th Report," 22 (emphasis added).

59. Rabinowitz et al., "Rural and Underserved Areas," 256–259.

60. Rabinowitz et al., "Rural and Underserved Areas," 256–259.

61. Note that this is the way the Supreme Court has allowed race to be used in admissions decisions, rejecting a quota system. However, opponents of affirmative action often point out that because there are only a fixed number of slots available, these policies end up excluding some people on the basis of race even as they admit others on the same grounds. See *Grutter v. Bollinger*, 539 U.S. 306 (2003); *Gratz v. Bollinger*, 539 U.S. 244 (2003).

62. See, for example, Mark R. Wicclair, "Reasons and Health Care Professionals' Claims of Conscience," *Am. J. Bioethics* 7, no. 6 (2007): 21; Piers Benn, "The Role of Conscience in Medical Ethics," in Nafiska Athanssoulis, ed., *Philosophical Reflections on Medical Ethics* (New York: Palgrave Macmillan, 2005), 177. This issue was discussed in chapter 6.

63. Michelle Ko, Ronald A. Edelstein, Kevin C. Heslin, Shobita Rajagopalan, LuAnn Wilkerson, Lois Colburn, and Kevin Grumbach, "Impact of the University of California, Los Angeles/Charles R. Drew University Medical Education Program on Medical Students' Intentions to Practice in Underserved Areas," *Academic Med.* 80, no. 9 (2005): 803, 803 (discussing programs having the most success in producing rural physicians).

64. U.S. Government Accountability Office, "Health Professional Shortage Areas," 23.

65. Leonard D. Baer, Thomas C. Ricketts, Thomas R. Konrad, and Stephen S. Mick, "Do International Medical Graduates Reduce Rural Physician Shortages?," *Med. Care* 36, no. 11 (1998): 1534. See also L. Gary Hart, Susan M. Skillman, Meredith Fordyce, Matthew Thompson, Amy Hagopian, and Thomas

R. Konrad, "International Medical Graduate Physicians in the United States: Changes Since 1981," *Health Affairs* 26, no. 4 (2007): 1159.

66. For a description of this problem, see Sabina Alkire and Lincoln Chen, Joint Learning Initiative: Human Resources for Health and Development, "'Medical Exceptionalism' in International Migration: Should Doctors and Nurses Be Treated Differently?," as well as Dela Dovlo and Tim Martineau, Joint Learning Initiative: Human Resources for Health and Development, "A Review of the Migration of Africa's Health Professionals," both at www.globalhealthtrust.org.

67. Institute of Medicine Committee on Evaluating Clinical Applications of Telemedicine, *Telemedicine: A Guide to Assessing Telecommunications in Health Care*, ed. Marilyn J. Field (Washington, DC: National Academies Press, 1996), 1.

68. Council on Graduate Medical Education, "10th Report," 24–25.

69. Alex Dominguez, "Robot Visits Patients When Doctor Can't," *Washington Post*, July 13, 2007.

70. Norimitsu Onishi, "In Japan's Rural Areas, Remote Obstetrics Fills the Gap," *New York Times*, April 8, 2007.

71. Glenn W. Wachter, "Interstate Licensure of Telemedicine Practitioners," *Telemedicine Information Exchange*, March 10, 2000, http://tie.telemed.org.

72. Council on Graduate Medical Education, "10th Report," 15.

73. National Abortion Federation, "Fact Sheet on Certified Nurse-Midwives, Nurse Practitioners, and Physician Assistants as Abortion Providers," 2005, www.prochoice.org.

74. 448 U.S. 297 (1980).

75. 415 U.S. 361 (1973).

76. 415 U.S. at 385. For the same reason, the Court also refused to find that the benefits program was invidiously discriminatory. 415 U.S. at 383.

77. When the Supreme Court held in *Sherbert v. Verner*, 374 U.S. 398 (1963), that a state could not deny unemployment benefits to a Seventh-Day Adventist who had been fired because she would not work on Sundays, its holding was not based solely on the fact that the state's policy imposed an incidental burden on the free exercise of religion. Instead, its conclusion was based on the fact that the incidental burden could not be justified by any compelling government interest. *Employment Division v. Smith*, 494 U.S. 872 (1990), on the other hand, effectively overruled *Sherbert* by holding that no compelling government interest is necessary for neutral laws of general applicability that just so happen to burden free exercise rights. These laws are permissible so long as the government's interest is legitimate. For more detail, see the discussion of this issue in chapter 1. But the important thing to note here is that not all government benefits are actually meant to be incentives—for example, unemployment compensation is not meant to incentivize people not to work. Thus even when *Sherbert* was still controlling, government incentives were subject to the sort of less stringent analysis we see applied in *Johnson*.

78. The veterans' benefits can be cast in precisely the same way—payment for an extra service that the conscientious objector did not provide.

79. For example, 19 U.S.C. § 267 permits extra compensation for customs officials working on Sundays and holidays, even though certain religious believers could not take advantage of such bonuses. Similarly, in a case involving a challenge to territorial abortion regulations in Guam, a relatively small and predominantly Catholic community generally opposed to abortion, plaintiffs were awarded attorney's fees of double the lodestar amount based on the undesirability of the case and the likelihood that no other attorney on the island would have accepted it. Such fees were available only to the lawyer willing to provide an unpopular service, without regard to the reasons that he was willing and other attorneys were not. Teresa Stanton Collett, "The Common Good and the Duty to Represent: Must the Last Lawyer in Town Take Any Case?," *S. Tex. L. Rev.* 40 (1999): 137, 143 (discussing *Guam Society of Obstetricians and Gynecologists v. Ada*, 100 F.3d 691 (9th Cir. 1996)).

80. Jonathan Weisman, "Where Did the Doctor Go?," *Washington Post*, May 20, 2007, B2.

81. See John K. Inglehart, "Grassroots Activism and the Pursuit of an Expanded Physician Supply," *New Eng. J. Med.* 358, no. 16 (2008): 1741.

82. "Shortages are, by definition, relative, and what constitutes adequate service is highly dependent on subjective criteria." Council on Graduate Medical Education, "10th Report," 24. See also Inglehart, "Grassroots Activism," 1744.

83. Goodman et al., "Benchmarking the U.S. Physician Workforce."

84. Goodman et al., "Benchmarking the U.S. Physician Workforce."

85. The Health Systems Performance Assessment is a method used by WHO to measure and compare how well health systems with the same level of resources attain a variety of health-related goals. See www.who.int/health-systems-performance. Daniel Wikler has suggested that this performance assessment be used to establish the moral baseline of services patients should get, which in turn should be used to determine the standard of care that must be satisfied for subjects in the control groups of clinical trials. Daniel Wikler, Harvard Program in Ethics and Health Presentation, April 17, 2007. However, it could be similarly useful to determine the moral baseline of services that patients should have access to despite the allowance of conscientious refusals by physicians.

Chapter 8

1. Norman Daniels, *Just Health: A Population View* (Cambridge: Cambridge University Press, 2007), 254.

2. Many reports of physician supply concede that nationally there is an adequate number of doctors, but shortages exist in some locales and specialties. John K. Inglehart, "Grassroots Activism and the Pursuit of an Expanded Physician

Supply," *New Eng. J. Med.* 358, no. 16 (2008): 1741, 1742. In fact, regional supply of physicians varies by more than 50 percent, and between 1979 and 1999, four out of five new doctors settled in regions where the supply was already high. David C. Goodman and Elliott S. Fisher, "Physician Workforce Crisis? Wrong Diagnosis, Wrong Prescription," *New Eng. J. Med.* 358, no. 16 (2008): 1658, 1658, 1660.

3. In 2005, Dennis Cauchon reported that the country needs to train as many as 10,000 additional physicians each year to meet the growing needs of patients. Because it takes approximately a decade to train a doctor, he estimated that we will face a shortage of as many as 200,000 doctors by 2020 unless action to reverse the shortage is taken promptly. Dennis Cauchon, "Medical Miscalculation Creates Doctor Shortage," *USA Today*, March 2, 2005, www.usatoday.com. A 2007 *JAMA* article suggests that these concerns have not gone away. Mike Mitka, "Looming Shortage of Physicians Raises Concerns About Access to Care," *J. Am. Med. Ass'n* 297, no. 10 (2007): 1045, 1046. However, whether there truly is a physician shortage is open to significant debate. See Inglehart, "Grassroots Activism," and Goodman and Fisher, "Physician Workforce Crisis?"

4. Teresa Stanton Collett notes these arguments, but rejects them, in the legal context. "The Common Good and the Duty to Represent: Must the Last Lawyer in Town Take Any Case?," *S. Tex. L. Rev.* 40 (1999): 137, 164.

5. Howard Brody and Susan S. Night, "The Pharmacist's Personal and Professional Integrity," *Am. J. Bioethics* 7, no. 6 (2007): 16, 16.

6. Note that this is the critical distinction away from the patient-centric model of medical professionalism described in chapter 2. Proponents of that approach seem to argue that the physician is violating her duty to patients whenever she conscientiously objects, though Savulescu at least is willing to excuse such a violation of duty in some circumstances. Actually, however, the physician is only violating her duty when she conscientiously objects in the hard case (and also when she fails to satisfy the baseline obligations described in the next chapter).

7. Mary K. Collins, "Conscience Clauses and Oral Contraceptives: Conscientious Objection or Calculated Obstruction?," *Annals Health L.* 15 (2006): 37, 57.

8. Robert K. Vischer, "Heretics in the Temple of Law: The Promise and Peril of the Religious Lawyering Movement," *J.L. & Rel.* 19 (2004): 427, 461–462.

9. Karen E. Adams, "Moral Diversity among Physicians and Conscientious Refusal of Care in the Provision of Abortion Services," *J. Am. Med. Women's Ass'n* 58, no. 4 (2003): 223, 224. Note that this is the sort of test of validity of a physician's grounds for refusal that was rejected in chapter 6.

10. Elizabeth Fenton and Loren Lomasky, "Dispensing with Liberty: Conscientious Refusal and the 'Morning-After Pill,'" *J. Med. & Phil.* 30, no. 6 (2005): 579, 585. Their argument applies with equal cogency to physicians. See p. 591 n. 6.

11. Fenton and Lomasky, "Dispensing with Liberty," 585.

12. Robert E. Goodin, *Protecting the Vulnerable: A Reanalysis of Our Social Responsibilities* (Chicago: University of Chicago Press, 1985), 118.

13. Goodin, *Protecting the Vulnerable*, 67. While others may argue that the patient is vulnerable to the circumstances and not necessarily vulnerable to the doctor, the doctor has a duty to mitigate the patient's vulnerability to those circumstances.

14. 192 Cal. Rptr. 650 (Cal. Ct. App. 1983).

15. James F. Childress, "Appeals to Conscience," *Ethics* 89, no. 4 (1979): 315, 333.

16. Recall that failure to deal with this issue was the greatest problem facing Pellegrino's physician-centric model of medical professionalism.

17. Leslie Cannold, "Consequences for Patients of Health Care Professionals' Conscientious Actions: The Ban on Abortions in South Australia," *J. Med. Ethics* 20, no. 2 (1994): 80, 83–85.

18. Julian Savulescu, "Conscientious Objection in Medicine," *Brit. Med. J.* 332 (2006): 294, 296.

19. Refusals for financial or other pragmatic reasons, such as triage of existing patients, may also be permissible, but are beyond our scope here.

20. Martha Swartz, "Conscience Clauses or Unconscionable Clauses: Personal Beliefs Versus Professional Responsibilities," *Yale J. Health Pol'y, L., & Ethics* 6 (2006): 269, 277.

21. See, for example, *Warthen v. Toms River Comm. Mem. Hosp.*, 488 A.2d 299 (N.J. Super. Ct. App. Div. 1985); *Pierce v. Ortho Pharm. Corp.*, 417 A.2d 505 (N.J. 1980) (distinguishing refusals based on professional ethics, which might be an appropriate public policy reason for refusal, from refusals based on personal morality, which the courts rejected).

22. See the discussion of this issue in chapter 1.

23. See American College of Obstetricians and Gynecologists, *Ethics in Obstetrics and Gynecology*, 2nd ed. (Washington, DC: American College of Obstetricians and Gynecologists, 2004), 25, 61, www.acog.org; President's Commission for the Study of Ethical Problems in Medicine and Biomedical and Behavioral Research, *Making Health Care Decisions: The Ethical and Legal Implications of Informed Consent in the Patient-Practitioner Relationship*, vol. 1 (Washington, DC: Government Printing Office, October 1982), 3, www.bioethics.gov.

24. Unfortunately, however, even the statutes that rely on these terms often do a poor job defining them, if they define them at all. For example, see Me. Rev. Stat. Ann. Tit. 18-A, § 5-807 (2005) ("medically ineffective," "contrary to applicable health-care standards"); Md. Code Ann., Health-Gen. § 5-611(b)(1) (West 2005) ("medically ineffective"); N.D. Cent. Code § 23-06.5-09(3) (2005) ("reasonable medical standards"); Wyo. Stat. Ann. § 35-22-410(v) (2005) ("good faith medical judgment").

25. Rhodes, "Ethical Standard of Care," 77–78; Thomas May, "Rights of Conscience in Health Care," *Social Theory & Practice* 27 (2001): 111, 120; David Orentlicher, "Trends in Health Care Decisionmaking: The Limitations of Legislation," *Md. L. Rev.* 53 (1994): 1255, 1294.

26. Ronald Dworkin, "Who Should Shape Our Culture?," *L. Sch. Mag.* 15 (2005): 20, 21.

27. See Richard B. Gallagher, Timothy M. Hall, Gary A. Hughes, Steven D. Najarian, Jeffrey A. Schafer, and Jeffrey J. Shampo, "Employment Contracts, in General," 42 Am. Jur. 2d Injunctions § 127 (2006).

28. Jill C. Morrison, senior counsel, National Women's Law Center, "Dr. No? The Debate on Conscience in Health Care," Panel Discussion Hosted by the Pew Forum on Religion and Public Life, the American Constitution Society, and the Federalist Society, September 8, 2006, http://pewforum.org.

29. Health and Ethics Policies of the AMA House of Delegates, Policy 295.896. Dickens, Cook, and Wardle also suggest that physicians should at least receive training in all procedures within their chosen specialty. B. M. Dickens and R. J. Cook, "The Scope and Limits of Conscientious Objection," *Int'l J. Gynecology & Obstetrics* 71, no. 1 (2000): 71, 76; Lynn D. Wardle, "Protecting the Rights of Conscience of Health Care Providers," *J. Legal Med.* 14 (1993): 177, 221.

30. Tom C. W. Lin makes a very similar point as a matter of labor economics in his "Treating an Unhealthy Conscience: A Prescription for Medical Conscience Clauses," *Vt. L. Rev.* 31 (2006): 105, 122: "If all medical students thinking about entering the area of female reproductive health were forced to perform procedures they found morally objectionable, surely a number of them would avoid that practice and enter another. In a broader sense, if all healthcare providers were mandated to perform procedures they found morally objectionable, a significant number of people considering the vocation might completely opt out of that labor market and enter another specialty."

31. Childress, "Appeals to Conscience," 334. But note that this is essentially the constitutional compelling interest standard rejected by *Employment Division v. Smith* for free exercise challenges.

32. This conclusion stands regardless of whether the hard case stems from a maldistribution or true shortage of willing physicians; it is too dangerous to enforce a duty to provide care in either circumstance.

33. Of course, if she could not access the service even in that broader area, she would have a claim against the board.

34. Some who are unwilling to accept this gap in compensation may actually prefer a system of true strict liability for licensing boards on the grounds that even though licensing boards may not have failed in their obligations, and even though there really are no willing physicians available for recruitment at the moment, it is better to be unfair to the licensing board than to the patient whose access interests were inappropriately burdened as a result of conscientious refusal.

However, this could become quite complex because it would involve a determination of where these patients should have had access as an abstract matter based on the number of physicians that should be available despite conscientious refusal but nevertheless accounting for shortages stemming from non-conscience-based sources. These difficulties suggest that while it may be ideal to avoid the compensation gap, incremental resolution of the problems raised by conscientious refusal is, at the very least, an acceptable approach.

35. For example, the proposed Access to Legal Pharmaceuticals Act, S. 809, 109th Cong. (2005), provides that a pharmacy violating its obligations to fill legal prescriptions will be liable to the United States "for a civil penalty in an amount not exceeding $5,000 per day of violation, and not to exceed $500,000 for all violations adjudicated in a single proceeding."

36. To permit a cause of action against a state entity (here, the licensing board), the state must waive its sovereign immunity, which it can do specifically for this sort of suit, if it has not already adopted a more general waiver.

37. Boards might also decide to pay the compensation award even if it is more expensive than recruiting willing physicians if the political climate in a given area is such that the board does not want to ensure that a service is actually available, such as abortion. This would be troublesome, since access is the ideal, but at the very least, the provision of an award to patients denied services will give the unserved patient the wherewithal to travel to a location where the service is available. Therefore, this approach is still preferable to the current system for dealing with physician refusals.

38. Julie D. Cantor and Ken Baum, "The Limits of Conscientious Objection— May Pharmacists Refuse to Fill Prescriptions for Emergency Contraception?," *New Eng. J. Med.* 351, no. 19 (2004): 2008, 2012.

Chapter 9

1. Daniel E. Hall and Farr Curlin, "Can Physicians' Care Be Neutral Regarding Religion?," *Acad. Med.* 79, no. 7 (2004): 677, 679.

2. Robert M. Veatch, *The Patient-Physician Relation; The Patient as Partner, Part 2* (Bloomington: Indiana University Press, 1991), 150.

3. Edmund D. Pellegrino, "Patient and Physician Autonomy: Conflicting Rights and Obligations in the Physician-Patient Relationship," *J. Contemp. Health L. & Pol'y* 10 (1993): 47, 47 . Pellegrino also describes the doctor patient relationship as "one of mutual obligation—like any truly ethical relationship." P. 51.

4. Candace Cummins Gauthier, "The Virtue of Moral Responsibility and the Obligations of Patients," *J. Med. and Phil.* 30, no. 2 (2005): 153, 154.

5. See, for example, Gauthier, "The Virtue of Moral Responsibility," 158 ("while the exercise of autonomy, without consideration of the impact of one's choices on others is certainly possible and even understandable, it is also morally wrong in the sense of being morally irresponsible").

6. Daniel K. Sokol, "Virulent Epidemics and Scope of Healthcare Workers' Duty of Care," *Emerging Infectious Diseases* 12, no. 8 (2006): 1238, 1239.

7. Teresa Stanton Collett, "Protecting the Health Care Provider's Right of Conscience," Center for Bioethics and Human Dignity, 2004, www.consciencelaws .org. But see the discussion of emergency obligations later in this chapter.

8. David Orentlicher, "Trends in Health Care Decisionmaking; The Limits of Legislation," *Md. L. Rev.* 53 (1994): 1255, 1304–1305.

9. Sabrina Rubin Erdely, "Doctors' Beliefs Hinder Patient Care," *Self Magazine*, June 2007, www.self.com.

10. Orentlicher, "Trends in Health Care Decisionmaking," 1304–1305.

11. Rebecca Dresser, "Professionals, Conformity, and Conscience," *Hastings Ctr. Rep.* 35, no. 6 (2005): 9, 9.

12. Bernard M. Dickens, "Ethical Misconduct by Abuse of Conscientious Objection Laws," *Med. & L.* 25, no. 3 (2006): 513, 517.

13. Natalie Langlois, Note "Life-Sustaining Treatment Law: A Model for Balancing a Woman's Reproductive Rights with a Pharmacist's Conscientious Objection," *B.C. L. Rev.* 47 (2006): 815, 852.

14. See, for example, *In re Jobes*, 529 A.2d 434 (N.J. 1987).

15. 464 F.2d 772 (D.C. Cir. 1972).

16. President's Commission for the Study of Ethical Problems in Medicine and Biomedical and Behavioral Research, *Making Health Care Decisions: A Report on the Ethical and Legal Implications of Informed Consent in the Patient-Practitioner Relationship*, vol. 1 (October 1982), 76, www.bioethics.gov.

17. Adam Sonfield, "Rights vs. Responsibilities: Professional Standards and Provider Refusals," *The Guttmacher Report on Public Policy*, 2005, 8, www .guttmacher.org (quoting the AMA position statement on informed consent).

18. Dresser, "Conformity," 10; Laura D. Briley, "A Physician's Professional Duty to Inform Despite Personal Ethical Objections," *Current Surgery* 60, no. 6 (2003): 594, 594.

19. 18 Pa. Cons. Stat. § 3202(d).

20. See "Verdicts and Settlements: Treatment of Pregnancy Complications Yields $1.7 Million Med Mal Verdict," 27 *Pa. L. Wkly.* 1256, November 11, 2004 (discussing the unreported case of *Thomas v. Abdul-Malek*, No. 02-1374 (W.D. Pa July 29, 2004)). See also *Harbeson v. Parke-Davis, Inc.*, 656 P.2d 483, 491 (Wash. 1983) (en banc) (holding that health-care workers must give their patients all material information so they can make informed choices, but that such a duty "does not, however, affect in any way the right of a physician to refuse on moral or religious grounds to perform an abortion").

21. Miss. Code Ann. § 41-107-5.

22. Miss. Code Ann. § 41-107-3(f).

23. *Sard v. Hardy*, 281 Md. 432 (1977); *Cobbs v. Grant*, 8 Cal.3d 229 (1972). See also President's Commission, *Making Health Care Decisions*, 38.

24. "Physicians have a right to shun practices they judge immoral, but they have no right to withhold important information from their patients." "Doctors Who Fail Their Patients," editorial, *New York Times*, February 13, 2007, www .nytimes.com.

25. See, for example, *Kenny v. Wepman*, 753 A.2d 924 (R.I. 2000); *Matthies v. Mastromonaco*, 160 N.J. 26 (1999); *Martin by Scoptur v. Richards*, 192 Wis. 2d 156 (1995). Courts have been divided with regard to how the adequacy of informed consent should be measured; some ask what a reasonable physician would disclose, some ask what a reasonable patient would want to know, and some split the difference by examining both perspectives. While it may be more difficult to assess what a reasonable patient might want to know than it is to determine what reasonable physicians disclose, as has been noted by several courts, the patient's perspective is of paramount importance and is the most appropriate measure of full informed consent. See Laura Hunter Dietz, Alan J. Jacobs, Theresa L. Leming, and John R. Kennel, 61 Am. Jur. 2d Physicians, Surgeons, and Other Healers §§ 172–177 (2006).

26. Laura Hunter Dietz, Alan J. Jacobs, Theresa L. Leming, and John R. Kennel, 61 Am. Jur. 2d Physicians, Surgeons, and Other Healers § 174.

27. Note, however, the New York permits nondisclosure of risks that are "too commonly known to warrant disclosure," even if this particular patient may not have been aware of them. N.Y. Pub. Health Law § 2805-d(4)(a) (2006).

28. For example, the New York statute provides a defense to a malpractice claim based on lack of informed consent if the patient assured the physician that he "did not want to be informed of the matters to which he would be entitled to be informed." N.Y. Pub. Health Law § 2805-d(4)(b) (2006).

29. Pellegrino, "Patient and Physician Autonomy," 54–55. See also Julian Savulescu, "Conscientious Objection in Medicine," *Brit. Med. J.* 332 (2006): 294, 296 ("for patients to give valid consent to treatment, they must be informed of relevant alternatives and their risks and benefits (in a reasonable, complete, and unbiased way)").

30. John M. Thorpe and Watson A. Bowes, "Prolife Perinatologist—Paradox or Possibility?," *New Eng. J. Med.* 326, no. 18 (1992): 1217, 1219; Bruce A. Green, "The Role of Personal Values in Professional Decisionmaking," *Geo. J. Legal Ethics* 11 (1999): 19, 45.

31. Joan H. Krause, "The Brief Life of the Gag Clause: Why Anti-Gag Clause Legislation Isn't Enough," *Tenn. L. Rev.* 67, no. 1 (1999): 1, 2.

32. American Civil Liberties Union, Reproductive Freedom Project, "Religious Refusals and Reproductive Rights," 2002, 10, www.aclu.org. See also Leslie Cannold, "Consequences for Patients of Health Care Professionals' Conscientious Actions: The Ban on Abortions in South Australia," *J. Med. Ethics* 20, no.

2 (1994): 80, 82; Lynn D. Wardle, "Protecting the Rights of Conscience of Health Care Providers," *J. Legal Med.* 14 (1993): 177, 195.

33. American Medical Association, "Code of Medical Ethics," § 10.05.

34. See, for example, 745 Ill. Comp. Stat. 70/6. (2006); Okla. Stat. Ann. Tit. 63 § 1-741 (B) (West 2005); Nev. Rev. Stat. Ann. §§ 449.191 and 632.475. Recall, however, that absent an existing doctor-patient relationship, physicians generally have no legal obligation to assist an individual facing a medical emergency.

35. 42 U.S.C. § 1395dd.

36. Laura Hunter Dietz, Alan Jacobs, Theresa L. Leming, and John R. Kennel, 61 Am. Jur. 2d Physicians, Surgeons, and Other Healers §10 (2006).

37. See Alison McIntyre, "Doctrine of Double Effect," in Edward N. Zalta, ed., *The Stanford Encyclopedia of Philosophy*, summer 2006 ed., http://plato .stanford.edu, for an excellent primer on this theory. Also see Robert Young, "Voluntary Euthanasia," from the same source.

38. As McIntrye has explained, "to invoke double effect . . . is to make a comparative judgment: it is to assert that a harm that might permissibly be brought about as a *side effect* in promoting a good end could not permissibly be brought about as a *means* to the same good end." McIntyre, "Double Effect" (emphasis added). This distinction between means and side effects is one of the primary difficulties in proper application of the doctrine.

39. Note, however, that as we have seen previously, physicians currently have no legal duties to those who are not their patients. This includes emergency duties. Laura Hunter Dietz, Alan J. Jacobs, Theresa L. Leming, and John R. Kennel, "Obligation to Practice or Accept Professional Employment," 61 Am Jur 2d Physicians, Surgeons, and Other Healers § 121 (2007).

40. 42 U.S.C. § 1395dd(e).

41. Erdely, "Doctors' Beliefs Hinder Patient Care."

42. See, for example, Health and Ethics Policies of the AMA House of Delegates, Policy 130.970, "Access to Emergency Services."

43. Sonfield, "Rights vs. Responsibilities," 8 (quoting the World Medical Association Declaration on the Rights of the Patient). Note that Britain, France, Italy, Norway, Denmark, and the Netherlands, for example, all allow refusal but impose a legal obligation of referral. Louis-Jacques Van Bogaert, "The Limits on Conscientious Objection to Abortion in the Developing World," *Developing World Bioethics* 2, no. 2 (2002): 131, 142.

44. Katherine A. White, Note, "Crisis of Conscience: Reconciling Religious Health Care Providers' Beliefs and Patient Rights," *Stan. L. Rev.* 51 (1999): 1703, 1745–1746.

45. American Civil Liberties Union, Reproductive Freedom Project, "Religious Refusals," 2 (noting that many patients were left without any access to family planning services because they did not know how to get care outside of their plan).

46. Eric M. Levine, "A New Predicament for Physicians: The Concept of Medical Futility, the Physician's Obligation to Render Inappropriate Treatment, and the Interplay of the Medical Standard of Care," *J.L. & Health* 9 (1995): 69, 87–88; C. T. Drechsler, "Liability of Physician Who Abandons Case," 57 A.L.R.2d 432, § 2 (2006).

47. Rob Stein, "Pharmacists' Rights at Front of New Debate," *Washington Post*, March 28, 2005.

48. Pellegrino, "Patient and Physician Autonomy," 63.

49. Edmund D. Pellegrino, "Commentary: Value Neutrality, Moral Integrity, and the Physician," *J. L. Med. & Ethics* 28, no. 1 (2000): 78, 79. Note also that in explaining its Consensus Statement on the Ethic of Medicine, the Council of Medical Specialty Societies states that "in respecting the patient's wishes, autonomy, and values, the physician is under no obligation to refer a patient explicitly for what he would assess to be an unethical procedure ... [although the physician is] free to provide the patient with a list of other physicians for treatment." Principle 4 of the Consensus Statement provides that physicians have a responsibility to maintain moral integrity. Sara C. Charles, Jeremy A. Lazarus, and Workgroup on the Ethic of Medicine, *Reframing the Professional Ethic: The Council of Medical Specialty Societies Consensus Statement on the Ethic of Medicine, West J. Med* 173 (2000): 198, 200.

50. Edmund D. Pellegrino, "The Physician's Conscience, Conscience Clauses, and Religious Belief: A Catholic Perspective," *Fordham Urb. L.J.* 30 (2002): 221, 239–240.

51. Pellegrino, "Commentary," 79.

52. Pellegrino, "The Physician's Conscience, Conscience Clauses, and Religious Belief," 239–240.

53. President's Commission, *Making Health Care Decisions*, 76.

54. American Medical Association, Medical Student Section, "Physician Objection to Treatment and Individual Patient Discrimination," Referred to Reference Committee on Amendments to Constitution and Bylaws, AMA House of Delegates, Received May 2, 2006, www.ama-assn.org.

55. Robert M. Sade, American Medical Association, "Report of the Council on Ethical and Judicial Affairs 6-A-07: Physician Objection to Treatment and Individual Patient Discrimination," referred to Committee on Amendments to Constitution and Bylaws (2007), http://www.ama-assn.org/ama1/pub/upload/mm/369/ceja_6a07.pdf. Rather than making any changes to existing AMA policy, however, CEJA interpreted existing policies as leading to this conclusion. It recognized that Principle VI creates freedom for physicians to choose their patients, but also noted that the conscientious refuser risks undermining other principles requiring responsibility to patients to be paramount (VII) and to support access to medical care (IX). Thus, a balance between these principles requires referral in most circumstances. CEJA's recommendation was simply to reaffirm the following policies: § 8.11, "Neglect of Patient," § 8.115 "Termination of the

Physician-Patient Relationship," § 9.06 "Free Choice," § 9.12 "Patient-Physician Relationship: Respect for Law and Human Rights," § 10.01, "Fundamental Elements of the Patient-Physician Relationship," and § 10.05, "Potential Patients."

56. American College of Obstetricians and Gynecologists, Committee Opinion, "The Limits of Conscientious Refusal in Reproductive Medicine," no. 385 (November 2007), www.acog.org.

57. Mike Leavitt, Secretary, Dept. of Health and Human Services, Letter to Norman F. Gant, Exec. Dir., Amer. Board of Obstetrics and Gynecology, March 14, 2008, www.hhs.gov.

58. Stephen F. Hanlon, American Bar Association Section of Individual Rights and Responsibilities, Health Law Section, Report to the House of Delegates, February 2005, 1, www.abanet.org (emphasis added).

59. Susan Berke Fogel and Lourdes A. Rivera, "Religious Beliefs and Healthcare Necessities: Can They Coexist?," *ABA Human Rights Magazine*, 2002, www.abanet.org (also arguing that health-care professionals must provide complete and accurate medical information, as well as urgent care).

60. Angela Bonavoglia, "Co-Opting Conscience: The Dangerous Evolution of Conscience Clauses in American Health Policy," Pro Choice Resource Center, 1999, 3, www.prochoiceresource.com; Sylvia A. Law, "Silent No More: Physicians' Legal and Ethical Obligations to Patients Seeking Abortions," *N.Y.U. Rev. L. & Soc. Change* 21 (1995): 279, 295; Martha Minow, "On Being a Religious Professional: The Religious Turn in Professional Ethics," *U. Pa. L. Rev.* 150 (2001): 661, 682; Dickens, "Ethical Misconduct," 521; Savulescu, "Conscientious Objection in Medicine," 296.

61. John K. Davis, "Conscientious Refusal and a Doctor's Right to Quit," *J. Med. & Phil.* 29, no. 1 (2004): 75, 90 n. 14.

62. Veatch, *Patient-Physician Relation*, 151.

63. See, for example, *Conservatorship of Morrison v. Abramovice*, 253 Cal.Rptr. 530, 534 (Ct. App. 1988).

64. Veatch, *Patient-Physician Relation*, 152. Veatch is stumped because he argues that surely "the physician should have the right not to refer for what he considers to be murder" and surely "the woman should have the right to speak with a physician who may condone legal abortion."

65. Note that this is precisely what Texas requires when patients and physicians or institutions disagree as to whether continued life-sustaining treatment is appropriate. Texas Advance Directive Act, §§ 166.046 and 166.052.

66. See I Corinthians 5: 9–10 ("I wrote you in my letter not to associate with immoral people, not at all referring to the immoral of this world or the greedy and robbers or idolaters; for you would then have to leave the world"); Christian Medical and Dental Associations, "Moral Complicity with Evil," 2004, www.cmawashington.org. But note that criminal conspiracy law often does not distinguish between the liability of the perpetrator and the accomplice, though it does consider knowledge and intent.

67. Orville N. Griese, *Catholic Identity in Health Care: Principles and Practice* (Braintree, MA: Pope John Center, 1987), 388–390.

68. Edmund D. Pellegrino, "Balancing Science, Ethics and Politics: Stem Cell Research, a Paradigm Case," *J. Contemp. Health L. & Pol'y* 18 (2002): 591, 603. While Pellegrino clearly believes that there is an insufficient moral distance separating referral from actual provision of the objectionable service, perhaps he would find the more attenuated assistance described here less troublesome. Howard Brody and Susan S. Night also recognize that referral and actual provision are morally different. "The Pharmacist's Personal and Professional Integrity," *Am. J. Bioethics* 7, no. 6 (2007): 16, 17.

69. Edmund D. Pellegrino, "Professionalism, Profession and the Virtues of the Good Physician," *Mt. Sinai J. Med.* 69, no. 6 (2002): 378, 382.

70. Martin Benjamin, "Conscience," in Stephen G. Post, ed., *The Encyclopedia of Bioethics*, 3rd ed. (New York: Macmillan Reference, 2004), 515.

71. Pellegrino, "The Physician's Conscience, Conscience Clauses, and Religious Belief," 243.

72. Jeffrey Blustein, "Doing What the Patient Orders: Maintaining Integrity in the Doctor-Patient Relationship," *Bioethics* 7, no. 4 (1993): 289, 300; Pellegrino, "Commentary," 80.

73. Dianne N. Irving, Catholic Medical Association of the U.S. and International Federation of Catholic Medical Associations, "Science, the Formation of Conscience and Moral Decision Making," www.consciencelaws.org ("any scientific error in the beginning of [the] moral decision making process ... precludes us from making morally correct decisions in the end").

74. Marie McCullough, "'Morning After' Pill Gets Final Approval," *Philadelphia Inquirer*, August 25, 2006, www.philly.com (noting that Plan B is not an abortion pill, since it has no effect on an established pregnancy, working instead by preventing ovulation, fertilization, or implantation).

75. Blustein, "Doing What the Patient Orders," 297.

76. Pellegrino, "Commentary," 80.

77. John Stuart Mill, "On Liberty," *The Basic Writings of John Stuart Mill* (New York: Modern Library, 2002), 79.

78. Pellegrino, "Commentary," 80. See also Russell G. Pearce, "The Legal Profession as a Blue State: Reflections on Public Philosophy, Jurisprudence, and Legal Ethics," *Fordham L. Rev.* 75 (2007): 1339, 1364 (discussing the need for lawyers to counsel clients on the moral implications of their choices).

79. Pellegrino, "Patient and Physician Autonomy," 55.

80. Howard Lesnick, "The Religious Lawyer in a Pluralist Society," *Fordham L. Rev.* 66 (1998): 1469, 1497.

81. Timothy E. Quill and Howard Brody, "Physician Recommendations and Patient Autonomy: Finding a Balance between Physician Power and Patient Choice," *Ann. Int. Med.* 125, no. 9 (1996): 763, 767.

82. The obligations described in this chapter have been recently espoused by ACOG:

Although respect for conscience is important, conscientious refusals should be limited if they constitute an imposition of religious or moral beliefs on patients, negatively affect a patient's health, are based on scientific misinformation, or create or reinforce racial or socioeconomic inequalities. Conscientious refusals that conflict with patient well-being should be accommodated only if the primary duty to the patient can be fulfilled. All health care providers must provide accurate and unbiased information so that patients can make informed decisions. Where conscience implores physicians to deviate from standard practices, they must provide potential patients with accurate and prior notice of their personal moral commitments. Physicians and other health care providers have the duty to refer patients in a timely manner to other providers if they do not feel that they can in conscience provide the standard reproductive services that patients request. In resource-poor areas, access to safe and legal reproductive services should be maintained. Providers with moral or religious objections should either practice in proximity to individuals who do not share their views or ensure that referral processes are in place. In an emergency in which referral is not possible or might negatively have an impact on a patient's physical or mental health, providers have an obligation to provide medically indicated and requested care. ACOG, Committee Opinion, No. 385, 1.

Chapter 10

1. For example, Thomas May has noted that the acceptability of referral as a way around conscientious refusal depends on whether the patient would be significantly less comfortable with the new provider. Thomas May, "Rights of Conscience in Health Care," *Social Theory & Practice* 27 (2001): 111, 126.

2. Laura D. Briley, "A Physician's Professional Duty to Inform Despite Personal Ethical Objections," *Current Surgery* 60, no. 6 (2003): 594, 595.

3. 517 A.2d 886, 889–890 (Super. Ct. N.J. 1986).

4. Robert M. Veatch and Carol Mason Spicer, "Medically Futile Care: The Role of the Physician in Setting Limits," *Am. J. L. & Med.* 18, no. 1-2 (1992): 15, 24.

5. *FTC v. Tenet Health Care Corp.*, 186 F.3d 1045, 1055 (8th Cir. 1999).

6. For example, a recent magazine story reports that one woman who was denied emergency contraception following her rape has not seen a gynecologist since, "for fear of being judged again." Sabrina Rubin Erdely, "Doctors' Beliefs Hinder Patient Care," *Self Magazine*, June 2007, www.self.com. This example, however, seems to present a paradigm case of refusal that was not carried out in an appropriate manner.

7. R. Alta Charo, "The Celestial Fire of Conscience—Refusing to Deliver Medical Care," *New Eng. J. Med.* 352, no. 24 (2005): 2471, 2471.

8. Notably, even if states do not promptly adopt this legislation creating consequences for state boards that fail to remedy access problems resulting from con-

scientious refusal, many licensing boards would still have the capacity to take this responsibility, if not compensation structure, on themselves. This is because most state medical practice acts are quite liberal in their delegation of power to licensing boards. The California statute provides, for example, that "protection of the public shall be the highest priority for the Medical Board of California in exercising its licensing, regulatory, and disciplinary functions. Whenever the protection of the public is inconsistent with other interests sought to be promoted, the protection of the public shall be paramount." Cal. Bus. and Prof. Code § 2001.1. Thus, it appears that the Medical Board could take measures to match supply and demand of its own accord given the argument that physician diversity and patient access are both in the public interest. Nevertheless, the funding issue is key and that will take legislative buy-in; in the meantime, however, these issues are sufficiently important that boards ought to do all they can on their own. At the very least, they ought to advocate for both patients and doctors by pressuring legislatures in all permissible ways to move forward with this proposal.

9. Adam Sonfield, "Rights vs. Responsibilities: Professional Standards and Provider Refusals," *The Guttmacher Report on Public Policy*, 2005, 8, www.guttmacher.org.

10. Sonfield, "Rights vs. Responsibilities," 8.

11. Judith F. Daar, "A Clash at the Bedside: Patient Autonomy v. a Physician's Professional Conscience," *Hastings L.J.* 44 (1993): 1241, 1275–1276.

12. Kent Greenawalt, "Objections in Conscience to Medical Procedures: Does Religion Make a Difference?," *U. Ill. L. Rev.* (2006): 799, 824.

13. 745 Ill. Comp. Stat. Ann. 70/4 (West 2006); Miss. Code Ann. § 41-107-5 (2005); Wash. Rev. Code § 70.47.160 (West 2006).

14. John W. Kingdon, *Agendas, Alternatives, and Public Policies*, 2nd ed. (New York: HarperCollins, 1995), 174–175.

15. Carol S. Weissert and William G. Weissert, *Governing Health: The Politics of Health Policy*, 2nd ed. (Baltimore: Johns Hopkins University Press, 2002), 313.

16. James F. Childress and Mark Siegler, "Metaphors and Models of Doctor-Patient Relationships: Their Implications for Autonomy," *Theor. Med. & Bioethics* 5, no. 1 (1984): 17, 26.

17. Martin Benjamin, "Conscience," in Stephen G. Post, ed., *The Encyclopedia of Bioethics*, 3rd ed. (New York: Macmillan Reference, 2004), 515.

18. Benjamin, "Conscience," 516. In fact, Benjamin argues that conscientious physicians are so valuable that health-care institutions have "prudential as well as ethical grounds for accommodating their claims of conscience even at the cost of some inconvenience or expense."

19. Elizabeth Fenton and Loren Lomasky, "Dispensing with Liberty: Conscientious Refusal and the 'Morning-After Pill,'" *J. Med. & Phil.* 30, no. 6 (2005): 579, 583.

20. Piers Benn, "The Role of Conscience in Medical Ethics," in Nafiska Athanssoulis, ed., *Philosophical Reflections on Medical Ethics* (New York: Palgrave Macmillan, 2005), 169; Farr A. Curlin, "Caution: Conscience Is the Limb on Which Medical Ethics Sits," *Am. J. Bioethics* 7, no. 6 (2007): 30, 31.

21. Sean Murphy, "Service or Servitude: Reflections on Freedom of Conscience for Health Care Workers," 1, www.consciencelaws.org.

22. John J. Hardt, "The Necessity of Conscience and the Unspoken Ends of Medicine," *Amer. J. Bioethics* 7, no. 6 (2007): 18 (emphasis added).

References

"Abortion Ultrasound Bill Advances in S.C.; First-of-Its-Kind Law Would Require Women to See Ultrasound Before Having Abortion." *CBS News,* March 21, 2007. www.cbsnews.com.

Adams, Karen E. "Moral Diversity among Physicians and Conscientious Refusal of Care in the Provision of Abortion Services." *J. Am. Med. Women's Ass'n* 58, no. 4 (2003): 223.

Alexander, John K. "Promising, Professional Obligations, and the Refusal to Provide Service." *HEC Forum* 17, no. 3 (2005): 178.

Alkire, Sabina, and Chen, Lincoln. "'Medical Exceptionalism' in International Migration: Should Doctors and Nurses Be Treated Differently?" Joint Learning Initiative: Human Resources for Health and Development. www.globalhealthtrust.org.

American Bar Association. "Public Perceptions of Lawyers: Consumer Research Findings." 2002. www.abanet.org.

American Civil Liberties Union, Reproductive Freedom Project. "Religious Refusals and Reproductive Rights." 2002. www.aclu.org.

American College of Obstetricians and Gynecologists. *Ethics in Obstetrics and Gynecology.* 2nd ed. Washington, DC: American College of Obstetricians and Gynecologists, 2004. www.acog.org.

American College of Obstetricians and Gynecologists, Committee on Gynecologic Practice and Committee on International Affairs. "Female Genital Mutilation." *Int. J. Gynecology & Obstetrics* 49, no. 2 (1995): 209.

American College of Obstetricians and Gynecologists. Committee Opinion, "The Limits of Conscientious Refusal in Reproductive Medicine." No. 385 (November 2007). www.acog.org.

American College of Obstetricians and Gynecologists, Executive Board. "ACOG Statement of Policy: Abortion Policy." 2004. www.acog.org.

American Medical Association. "Code of Medical Ethics." www.ama-assn.org.

American Medical Association. "Health and Ethics Policies of the AMA House of Delegates." www.ama-assn.org.

American Medical Association. "Principles of Medical Ethics." www.ama-assn
.org.

American Medical Association, Medical Student Section. "Physician Objection to Treatment and Individual Patient Discrimination." www.ama-assn.org.

American Society for Reproductive Medicine, Ethics Committee. "Access to Fertility Treatment by Gays, Lesbians, and Unmarried Persons." *Fertility & Sterility* 86, no. 5 (2006): 1333.

Appel, Jacob M. "May Doctors Refuse Infertility Treatments to Gay Patients?" *Hastings Ctr. Rep.* 36, no. 4 (2006): 20.

Applbaum, Arthur. "Doctor, Schmocter: Practice Positivism and Its Complications." In Robert B. Baker, Arthur L. Caplan, Linda L. Emanuel, and Stephen R. Latham, eds., *The American Medical Ethics Revolution*. Baltimore: Johns Hopkins University Press, 1999.

Arnold, Louise, and Stern, David Thomas. "What Is Medical Professionalism?" In David Stern, ed., *Measuring Medical Professionalism*. Oxford: Oxford University Press, 2006.

The "Ashley Treatment." http://ashleytreatment.spaces.live.com.

Baer, Leonard D., Ricketts, Thomas C., Konrad, Thomas R., and Mick, Stephen S. "Do International Medical Graduates Reduce Rural Physician Shortages?" *Med. Care* 36, no. 11 (1998): 1534.

Balint, John. "There Is a Duty to Treat Noncompliant Patients." *Seminars in Dialysis* 14, no. 1 (2001): 28.

Bassett, William W. "Private Religious Hospitals: Limitations upon Autonomous Moral Choices in Reproductive Medicine." *J. Contemp. Health L. & Pol'y* 17 (2001): 455.

Baylis, Françoise. "Expert Testimony by Persons Trained in Ethical Reasoning: The Case of Andrew Sawatzky." *J.L. Med & Ethics* 28 (2000): 224.

Beauchamp, Tom L., and Childress, James. *The Principles of Biomedical Ethics*. 5th ed. Oxford: Oxford University Press, 2001.

Benjamin, Martin. "Conscience." In Stephen G. Post, ed., *The Encyclopedia of Bioethics*. 3rd ed. New York: Macmillan Reference, 2004.

Benn, Piers. "The Role of Conscience in Medical Ethics." In Nafiska Athanssoulis, ed., *Philosophical Reflections on Medical Ethics*. New York: Palgrave Macmillan, 2005.

Benson, Iain T. "'Autonomy,' 'Justice,' and the Legal Requirement to Accommodate the Conscience and Religious Beliefs of Professionals in Health Care." 2001. www.consciencelaws.org.

Berlinger, Nancy. "Martin Luther at the Bedside." *Hastings Ctr. Rep.* 37, no. 2 (2007): 49.

Blackman, Dana E. "Refusal to Dispense Emergency Contraception in Washington State: An Act of Conscience or Unlawful Sex Discrimination?" *Mich L. Gender & L.* 14 (2007): 59.

Bleich, J. David. "The Physician as a Conscientious Objector." *Fordham Urb. L.J.* 30 (2002): 245.

Blustein, Jeffrey. "Doing What the Patient Orders: Maintaining Integrity in the Doctor-Patient Relationship." *Bioethics* 7, no. 4 (1993): 289.

Blustein, Jeffrey. "Infertility Treatments for Gay Parents?" Letter to the editor. *Hastings Ctr. Rep.* 36, no. 5 (2006): 6.

Blustein, Jeffrey, and Fleischman, Alan R. "The Pro-Life Maternal-Fetal Medicine Physician: A Problem of Integrity." *Hastings Ctr. Rep.* 25, no. 1 (1995): 22.

Boddiger, David. "AMA Hopes Streamlined Agenda Will Boost Membership." *Lancet* 366, no. 9490 (2005): 971.

Bonavoglia, Angela. "Co-Opting Conscience: The Dangerous Evolution of Conscience Clauses in American Health Policy." Pro Choice Resource Center. 1999. www.prochoiceresource.com.

Boulware, L. Ebony, Cooper, Lisa A., Ratner, Lloyd E., LaVeist, Thomas A., and Powe, Neil R. "Race and Trust in the Health Care System." *Public Health Rep.* 118, no. 4 (2003): 358.

Briley, Laura D. "A Physician's Professional Duty to Inform Despite Personal Ethical Objections." *Current Surgery* 60, no. 6 (2003): 594.

Broad, C. D. *Ethics and the History of Philosophy.* London: Routledge, 1952.

Brody, Howard, and Night, Susan S. "The Pharmacist's Personal and Professional Integrity." *Am. J. Bioethics* 7, no. 6 (2007): 16.

Browne, Alister, Dickenson, Brent, and Van Der Waal, Rena. "The Ethical Management of the Noncompliant Patient." *Cambridge Q. Healthcare Ethics* 12 (2003): 289.

Brushwood, David B. "Conscientious Objection and Abortifacient Drugs." *Clinical Therapeutics* 15, no. 1 (1993): 204.

Budziszewski, J. "Handling Issues of Conscience." 1999 Beatty Memorial Lecture, McGill University College of Education. www.consciencelaws.org.

Caldwell, Christopher. "A DHumb Idea at 25." *The Weekly Standard*, April 13, 1998.

Cannold, Leslie. "Consequences for Patients of Health Care Professionals' Conscientious Actions: The Ban on Abortions in South Australia." *J. Med. Ethics* 20, no. 2 (1994): 80.

Cantor, Julie D. "When an Adult Female Seeks Ritual Genital Alteration: Ethics, Law, and the Parameters of Participation." *Plastic & Reconstructive Surgery* 117, no. 4 (2006): 1158.

Cantor, Julie D., and Baum, Ken. "The Limits of Conscientious Objection—May Pharmacists Refuse to Fill Prescriptions for Emergency Contraception?" *New Eng. J. Med.* 351, no. 19 (2004): 2008.

Caplan, Arthur. "Media's Cooing over Sextuplets Is a Disservice." *MSNBC*, June 19, 2007. www.msnbc.com.

Capozzi, James D., and Rhodes, Rosamond. "Coping with Racism in a Patient." *J. Bone & Joint Surgery* 88, no. 11 (2006): 2543.

Capron, Alexander M. "In re Helga Wanglie." *Hastings Ctr. Rep.* 21, no. 5 (1991): 26.

Capron, Alexander M. "Professionalism and Professional Ethics." In Robert B. Baker, Arthur L. Caplan, Linda L. Emanuel, and Stephen R. Latham, eds., *The American Medical Ethics Revolution*. Baltimore: Johns Hopkins University Press, 1999.

Card, Robert. "Conscientious Objection and Emergency Contraception." *Am. J. Bioethics* 7, no. 6 (2007): 8.

Casavant, Vanessa Renée. "Catman's Transformation Raises Concerns over Extreme Surgery." *Seattle Times*, August 16, 2005.

Catholics for Free Choice. "Position Paper on Individual and Institutional Refusal Clauses." April 2005. www.catholicsforchoice.org.

Cauchon, Dennis. "Medical Miscalculation Creates Doctor Shortage." *USA Today*, March 2, 2005. www.usatoday.com.

Center on Conscience and War. "Conscientious Objectors and the Draft." 2001. www.centeronconscience.org.

Chandrasekhar, Charu A. "RX for Drugstore Discrimination: Challenging Pharmacy Refusals to Dispense Prescription Contraceptives under State Public Accommodation Laws." *Alb. L. Rev.* 70 (2006): 55.

Charatan, Fred. "US Doctors Debate Refusing Treatment to Malpractice Lawyers." *Brit. Med. J.* 328 (2004): 1518.

Charles, Sara C., Lazarus, Jeremy A., and Workgroup on the Ethic of Medicine. "Reframing the Professional Ethic: The Council of Medical Specialty Societies Consensus Statement on the Ethic of Medicine." *West J. Med.* 173 (2000): 198.

Charo, R. Alta. "The Celestial Fire of Conscience—Refusing to Deliver Medical Care." *New Eng. J. Med.* 352, no. 24 (2005): 2471.

Charo, R. Alta. "Health Care Provider Refusals to Treat, Prescribe, Refer, or Inform: Professionalism and Conscience." American Constitution Society, 2007. www.americanconstitutionsociety.org.

Childress, James F. "Appeals to Conscience." *Ethics* 89, no. 4 (1979): 315.

Childress, James F., and Siegler, Mark. "Metaphors and Models of Doctor-Patient Relationships: Their Implications for Autonomy." *Theor. Med.* 5, no. 1 (1984): 17.

Christian Medical & Dental Associations. "Moral Complicity with Evil." 2004. www.cmawashington.org.

Clark, Peter A. "Physician Participation in Executions: Care Giver or Executioner?" *J. L. Med. & Ethics* 34, no. 1 (2006): 95.

Coleman, Carl H. "Conceiving Harm: Disability Discrimination in Assisted Reproductive Technologies." *UCLA L. Rev.* 50 (2002): 17.

Collett, Teresa Stanton. "The Common Good and the Duty to Represent: Must the Last Lawyer in Town Take Any Case?" *S. Tex. L. Rev.* 40 (1999): 137.

Collett, Teresa Stanton. "Professional versus Moral Duty: Accepting Appointments in Unjust Civil Cases." *Wake Forest L. Rev.* 32 (1997): 635.

Collett, Teresa Stanton. "Protecting the Health Care Provider's Right of Conscience." Center for Bioethics and Human Dignity. 2004. www.consciencelaws .org.

Collins, Mary K. "Conscience Clauses and Oral Contraceptives: Conscientious Objection or Calculated Obstruction?" *Ann. Health L.* 15 (2006): 37.

Coulehan, Jack, Williams, Peter C., McCrary, S. Van, and Belling, Catherine. "The Best Lack All Conviction: Biomedical Ethics, Professionalism, and Social Responsibility." *Cambridge Q. Healthcare Ethics* 12, no. 1 (2003): 21.

Council on Graduate Medical Education. "10th Report: Physician Distribution and Health Care Challenges in Rural and Inner-City Areas." February 4, 1998. http://cogme.gov/10.pdf.

Council of the Society of Obstetricians and Gynecologists of Canada. "Gender Selection." www.csdms.com.

Crary, David. "Sales Soar for Morning-After Pill." Associated Press, August 22, 2007.

Crosby, John F. "There Is No Moral Authority in Medicine: Response to Cowdin and Tuohey." *Christian Bioethics* 4, no. 1 (1998): 63.

Crowley, Christopher. "A New Rejection of Moral Expertise." *Med., Health Care & Phil.* 8, no. 3 (2005): 273.

Cruess, Richard L., Cruess, Sylvia R., and Johnston, Sharon E. "Professionalism and Medicine's Social Contract." *J. Bone & Joint Surg.* 82-A (2000): 1189.

Cruess, Sylvia R., Johnston, Sharon E., and Cruess, Richard L. "Professionalism for Medicine: Opportunities and Obligations." *Iowa Orthopaedic J.* 24 (2004): 9.

Cunningham, Larry. "Can a Catholic Lawyer Represent a Minor Seeking a Judicial Bypass for an Abortion? A Moral and Canon Law Analysis." *J. Cath. Legal Stud.* 44, no. 2 (2005): 379.

Curlin, Farr A. "Caution: Conscience Is the Limb on Which Medical Ethics Sits." *Am. J. Bioethics* 7, no. 6 (2007): 30.

Curlin, Farr A., and Hall, Daniel E. "Strangers or Friends? A Proposal for a New Spirituality-in-Medicine Ethic." *J. Gen. Intern. Med.* 20, no. 4 (2005): 370.

Curlin, Farr A., Lawrence, Ryan E., Chin, Marshall H., and Lantos, John D. "Religion, Conscience, and Controversial Clinical Practices." *New Eng. J. Med.* 356, no. 6 (2007): 593.

Daar, Judith F. "A Clash at the Bedside: Patient Autonomy v. a Physician's Professional Conscience." *Hastings L.J.* 44 (1993): 1241.

Daniels, Norman. "Duty to Treat or Right to Refuse?" *Hastings Ctr. Rep.* 21, no. 2 (1991): 36.

Daniels, Norman. *Just Health: A Population View*. Cambridge: Cambridge University Press, 2007.

Danis, Marion, and Churchill, L. R. "Autonomy and the Common Weal." *Hastings Ctr. Rep.* 21, no. 1 (1991): 25.

Davis, John K. "Conscientious Refusal and a Doctor's Right to Quit." *J. Med. & Phil.* 29, no. 1 (2004): 75.

de Melo-Martin, Inmaculada. "Should Professional Associations Sanction Conscientious Refusals?" *Am. J. Bioethics* 7, no. 6 (2007): 23.

Dickens, Bernard M. "Ethical Misconduct by Abuse of Conscientious Objection Laws." *Med. & Law* 25, no. 3 (2006): 513.

Dickens, Bernard M. "Preimplantation Genetic Diagnosis and 'Saviour Siblings.'" *Int. J. Gynecology & Obstetrics* 88 (2005): 91.

Dickens, Bernard M., and Cook, Rebecca J. "The Scope and Limits of Conscientious Objection." *Int'l J. Gynecology & Obstetrics* 71, no. 1 (2000): 71.

Dietz, Laura Hunter, Jacobs, Alan J., Leming, Theresa L., and Kennel, John R. "Creation and Nature of Relationship." 61 Am. Jur. 2d Physicians, Surgeons, and Other Healers § 130 (2006).

Dietz, Laura Hunter, Jacobs, Alan J., Leming, Theresa L., and Kennel, John R. "Duty to Refer Patient to Specialist or Qualified Practitioner." 61 Am. Jur. 2d Physicians, Surgeons, and Other Healers § 214 (2006).

Dietz, Laura Hunter, Jacobs, Alan J., Leming, Theresa L., and Kennel, John R. "Obligation to Practice or Accept Professional Employment." 61 Am. Jur. 2d Physicians, Surgeons, and Other Healers § 121 (2007).

Dietz, Laura Hunter, Jacobs, Alan J., Leming, Theresa L., Martin, Lucas, Shampo, Jeffrey, Surette, Eric, and Zakolski, Lisa. "Moral or Humanitarian Considerations; Duty to Aid or Protect Others." 57A Am. Jur. 2d Negligence § 90 (2006).

Dominguez, Alex. "Robot Visits Patients When Doctor Can't." *Washington Post*, July 13, 2007.

Dovlo, Dela, and Martineau, Tim. "A Review of the Migration of Africa's Health Professionals." Joint Learning Initiative: Human Resources for Health and Development. www.globalhealthtrust.org.

Dowling, Katherine. "Prolife Doctors Should Have Choices, Too." *U.S. Catholic*, March 2001.

Drechsler, C. T. "Liability of Physician Who Abandons Case." 57 A.L.R.2d 432 (2007).

Dresser, Rebecca S. "Freedom of Conscience, Professional Responsibility, and Access to Abortion." In Belinda Bennett, ed., *Abortion*. Burlington, VT: Ashgate, 2004.

Dresser, Rebecca S. "Professionals, Conformity, and Conscience." *Hastings Ctr. Rep.* 35, no. 6 (2005): 9.

Dudzinski, Denise M. "Integrity in the Relationship between Medical Ethics and Professionalism." *Am. J. Bioethics* 4, no. 2 (2004): 26.

Dworkin, Ronald. *Taking Rights Seriously.* Cambridge: Harvard University Press, 2005.

Dworkin, Ronald. "Who Should Shape Our Culture?" *L. Sch. Mag.* 15 (2005): 20.

Egelko, Bob. "State High Court to Hear Lesbian's Case; Doctors Denied Her Infertility Treatment on Religious Grounds." *San Francisco Chronicle*, June 15, 2006. www.sfgate.com.

Eisgruber, Christopher L., and Sager, Lawrence G. "The Vulnerability of Conscience: The Constitutional Basis for Protecting Religious Conduct." *U. Chi. L. Rev.* 61 (1994): 1245.

Elhauge, Einer. "Allocating Health Care Morally." *Cal. L. Rev.* 82 (1994): 1449.

Elliott, Carl. *Better Than Well: American Medicine Meets the American Dream.* New York: Norton, 2003.

Emanuel, Ezekiel J., and Emanuel, Linda L. "Four Models of the Physician-Patient Relationship." *J. Amer. Med. Ass'n* 267, no. 16 (1992): 2221.

"Embryos Screened to Create 'Savior Siblings'; Babies Compatible as Stem-Cell Donors." *Milwaukee J. Sentinel*, May 4, 2004.

Erdely, Sabrina Rubin. "Doctors' Beliefs Hinder Patient Care." *Self Magazine*, June 2007. www.self.com.

Eskridge, William N., Jr., Frickey, Philip P., and Garrett, Elizabeth, eds. *Cases and Materials on Legislation: Statutes and the Creation of Public Policy*, 3rd ed. St. Paul, MN: West Publishing Co., 2001.

Federation of State Medical Boards. *A Guide to the Essentials of a Modern Medical Practice Act.* 10th ed. 2003. www.fsmb.org.

Federation of State Medical Boards. "Protecting the Public: How State Medical Boards Regulate and Discipline Physicians." www.fsmb.org.

Fenton, Elizabeth, and Lomasky, Loren. "Dispensing with Liberty: Conscientious Refusal and the 'Morning-After Pill.'" *J. Med. & Phil.* 30, no. 6 (2005): 579.

Floyd, Timothy. "Does Professionalism Leave Room for Religious Commitment?" *Fordham Urb. L.J.* 26 (1999): 875.

Fogel, Susan Berke, and Rivera, Lourdes A. "Religious Beliefs and Healthcare Necessities: Can They Coexist?" *ABA Human Rights Magazine*, 2002. www.abanet.org.

Fried, Charles. "The Lawyer as Friend: The Moral Foundations of the Lawyer-Client Relation." *Yale L.J.* 85 (1976): 1060.

Gallagher, Richard B., Hall, Timothy M., Hughes, Gary A., Najarian, Steven D., Schafer, Jeffrey A., and Shampo, Jeffrey J. "Employment Contracts, in General." 42 Am. Jur. 2d Injunctions § 127 (2006).

Gauthier, Candace Cummins. "The Virtue of Moral Responsibility and the Obligations of Patients." *J. Med. & Phil.* 30, no. 2 (2005): 153.

Gibbs, Nancy. "Defusing the War over the 'Promiscuity' Vaccine." *Time*, June 21, 2006. www.time.com.

Gibbs, Nancy. "Pillow Angel Ethics, Part 2." *Time*, January 9, 2007, 56.

Goodin, Robert E. *Protecting the Vulnerable: A Reanalysis of Our Social Responsibilities.* Chicago: University of Chicago Press, 1985.

Goodman, David C., and Fisher, Elliott S. "Physician Workforce Crisis? Wrong Diagnosis, Wrong Prescription." *New Eng. J. Med.* 358, no. 16 (2008): 1658.

Goodman, David C., Fisher, Elliott S., Bubolz, Thomas A., Mohr, Jack E., Posage, James F., and Wennberg, John E. "Benchmarking the U.S. Physician Workforce: An Alternative to Needs-Based or Demand-Based Planning." *J. Am. Med. Ass'n* 276, no. 22 (1996): 1811.

Gorovitz, Samuel. *Doctors' Dilemmas: Moral Conflict and Medical Care.* Oxford: Oxford University Press, 1982.

Gray, Bradley, and Stoddard, Jeffrey. "Patient-Physician Pairing: Does Racial and Ethnic Congruity Influence Selection of a Regular Physician?" *J. Cmty. Health* 22, no. 4 (1997): 47.

Green, Bruce A. "The Role of Personal Values in Professional Decisionmaking." *Geo. J. Legal Ethics* 11 (1999): 19.

Greenawalt, Kent. "Objections in Conscience to Medical Procedures: Does Religion Make a Difference?" *U. Ill. L. Rev.* (2006): 799.

Griese, Orville N. *Catholic Identity in Health Care: Principles and Practice.* Braintree, MA: Pope John Center, 1987.

"Growth Hormones for Kids?" *CBS News*, June 12, 2003. www.cbsnews.com.

Gurmankin, Andrea D., Caplan, Arthur L., and Braverman, Andrea M. "Screening Practices and Beliefs of Assisted Reproductive Technology Programs." *Fertility & Sterility* 83, no. 1 (2005): 61.

Guttmacher Institute. "State Policies in Brief: Refusing to Provide Health Services." 2006. www.guttmacher.org.

Guttmacher Institute. "State Policies in Brief: Restricting Insurance Coverage of Abortion." 2006. www.guttmacher.org.

Hall, Daniel E., and Curlin, Farr A. "Can Physicians' Care Be Neutral Regarding Religion?" *Acad. Med.* 79, no. 7 (2004): 677.

Hall, Mark A., Bobinski, Mary Anne, and Orentlicher, David. *Health Care Law and Ethics.* 6th ed. New York, NY: Aspen Publishers, 2003.

Hanlon, Stephen F. *ABA Section of Individual Rights and Responsibilities, Health Law Section, Report to the House of Delegates.* February 2005. www.abanet.org.

Hardt, John J. "The Necessity of Conscience and the Unspoken Ends of Medicine." *Amer. J. Bioethics* 7, no. 6 (2007): 18.

Hart, L. Gary, Skillman, Susan M., Fordyce, Meredith, Thompson, Matthew, Hagopian, Amy, and Konrad, Thomas R. "International Medical Graduate Physicians in the United States: Changes Since 1981." *Health Affairs* 26, no. 4 (2007): 1159.

Harvey, M. T. "What Does a 'Right' to Physician-Assisted Suicide (PAS) Legally Entail?" *Theoretical Med. & Bioethics* 23, no. 4–5 (2002): 271.

Hasson, Kevin Seamus. *The Right to Be Wrong: Ending the Culture War Over Religion in America*. San Francisco: Encounter Books, 2005.

"Influx of Doctors Overwhelms Texas Board." *Washington Post*, July 9, 2007.

Inglehart, John K. "Grassroots Activism and the Pursuit of Expanded Physician Supply." *New Eng. J. Med.* 358, no. 16 (2008): 1741.

Institute of Medicine, Committee on Evaluating Clinical Applications of Telemedicine. *Telemedicine: A Guide to Assessing Telecommunications in Health Care*. Ed. Marilyn J. Field. Washington, DC: National Academies Press, 1996.

Irving, Dianne N. "Science, the Formation of Conscience and Moral Decision Making." Catholic Medical Association of the U.S. and International Federation of Catholic Medical Associations. www.consciencelaws.org.

Jacobi, John V. "Consumer-Directed Health Care and the Chronically Ill." *U. Mich. J.L. Reform* 38 (2005): 531.

Jain, Tarun, Harlow, Bernard L., and Hornstein, Mark D. "Insurance Coverage and Outcomes of In Vitro Fertilization." *New Eng. J. Med.* 347, no. 9 (2002): 661.

Jansen, Lynn A. "No Safe Harbor: The Principle of Complicity and the Practice of Voluntary Stopping of Eating and Drinking." *J. Med. & Phil.* 29, no. 1 (2004): 61.

Jauhar, Sandeep. "A Patient's Demands versus a Doctor's Convictions." *New York Times*, April 3, 2007. www.nytimes.com.

Jones, James H. *Bad Blood: The Tuskegee Syphilis Experiment*. New York: Free Press, 1993.

Key, Frances L. "Female Circumcision/Female Genital Mutilation in the United States: Legislation and Its Implications for Health Providers." *J. Am. Med. Women's Ass'n* 52, no. 4 (1997): 179.

Kilner, John F. "Healthcare Resources, Allocation of; Macroallocation." In Stephen G. Post, ed., *The Encyclopedia of Bioethics*, 3rd ed. New York: Macmillan Reference, 2004.

King, David. "Why We Should Not Permit Embryos to Be Selected as Tissue Donors." *Bull. Med. Ethics* 190 (2003): 13.

Kingdon, John W. *Agendas, Alternatives, and Public Policies*, 2nd ed. New York: HarperCollins, 1995.

Ko, Michelle, Edelstein, Ronald A., Heslin, Kevin C., Rajagopalan, Shobita, Wilkerson, LuAnn, Colburn, Lois, and Grumbach, Kevin. "Impact of the University of California, Los Angeles/Charles R. Drew University Medical Education

Program on Medical Students' Intentions to Practice in Underserved Areas." *Academic Med.* 80, no. 9 (2005): 803.

Korn, Peter. "Doctor's Ethics Run Counter to Hospital Policy." *Portland Tribune*, April 21, 2006.

Krause, Joan H. "The Brief Life of the Gag Clause: Why Anti-Gag Clause Legislation Isn't Enough." *Tenn. L. Rev.* 67, no. 1 (1999): 1.

Krugman, Scott. "The Willowbrook Hepatitis Studies Revisited: Ethical Aspects." *Rev. Infectious Diseases* 8, no. 1 (1986): 157.

Kruse, Katherine R. "Lawyers, Justice, and the Challenge of Moral Pluralism." *Minn. L. Rev.* 90 (2005): 389.

Lambda Legal. "Benitez v. North Coast Women's Care Medical Group (Q & A)." www.lambdalegal.org.

Langlois, Natalie. Note, "Life-Sustaining Treatment Law: A Model for Balancing a Woman's Reproductive Rights with a Pharmacist's Conscientious Objection." *B.C. L. Rev.* 47 (2006): 815.

Laurance, Jeremy. "Abortion Crisis as Doctors Refuse to Perform Surgery." *The Independent*, April 16, 2007. http://news.independent.co.uk.

Law, Sylvia A. "Silent No More: Physicians' Legal and Ethical Obligations to Patients Seeking Abortions." *N.Y.U. Rev. L. & Soc. Change* 21 (1995): 279.

Lawrence, Ryan E., and Curlin, Farr A. "Clash of Definitions: Controversies about Conscience in Medicine." *Amer. J. Bioethics* 7, no. 12 (2007): 10.

Lesnick, Howard. "The Religious Lawyer in a Pluralist Society." *Fordham L. Rev.* 66 (1998): 1469.

Levine, Eric M. "A New Predicament for Physicians: The Concept of Medical Futility, the Physician's Obligation to Render Inappropriate Treatment, and the Interplay of the Medical Standard of Care." *J.L. & Health* 9 (1995): 69.

Levinson, Sanford. "Identifying the Jewish Lawyer: Reflections on the Construction of Professional Identity." *Cardozo L. Rev.* 14 (1993): 1577.

Lifton, Robert Jay. *The Nazi Doctors: Medical Killing and the Psychology of Genocide.* New York: Basic Books, 2000.

Lin, Tom C. W. "Treating an Unhealthy Conscience: A Prescription for Medical Conscience Clauses." *Vt. L. Rev.* 31 (2006): 105.

Loewy, Erich H. "Physicians and Patients: Moral Agency in a Pluralistic World." *J. Med. Human. & Bioethics* 7, no. 1 (1986): 57.

Manning, John F. "Lawmaking Made Easy." *Green Bag 2d* 10 (2007): 191.

Marshner, Connie. "The Health Insurance Exchange: Enabling Freedom of Conscience in Health Care." March 1, 2007. www.heritage.org.

Matthews, Richard S. "Indecent Medicine: In Defense of the Absolute Prohibition against Physician Participation in Torture." *Am. J. Bioethics* 6, no. 3 (2006): W34.

May, Thomas. "Rights of Conscience in Health Care." *Soc. Theory & Practice* 27, no. 1 (2001): 111.

McCann, Robert W. "Implications of the *Evanston Northwestern* Decision for Hospital Mergers and Health Care Antitrust Compliance." In *Health Law Handbook*. Minneapolis, MN: West, 2006.

McCarthy, Thomas R., and Thomas, Scott J. "Geographic Market Issues in Hospital Mergers." In *The Health Care Mergers and Acquisitions Handbook*. American Bar Association, 2003.

McCullough, Marie. "'Morning After' Pill Gets Final Approval." *Philadelphia Inquirer*, August 25, 2006. www.philly.com.

McIlheran, Patrick. "Don't Divorce Job from Conviction in Health Care." *Milwaukee Journal Sentinel*, February 13, 2005.

McIntyre, Alison. "Doctrine of Double Effect." In Edward N. Zalta, ed., *The Stanford Encyclopedia of Philosophy*. Summer 2006 ed. http://plato.stanford.edu.

Mechanic, David. "In My Chosen Doctor I Trust." *Brit. Med. J.* 329 (2004): 1418.

Meyers, Christopher, and Woods, Robert D. "Conscientious Objection? Yes, But Make Sure It Is Genuine." *Am. J. Bioethics* 7, no. 6 (2007): 19.

Meyers, Christopher, and Woods, Robert D. "An Obligation to Provide Abortion Services: What Happens When Physicians Refuse?" *J. Med. Ethics* 22, no. 2 (1996): 115.

Mill, John Stuart. "On Liberty." In *The Basic Writings of John Stuart Mill*. New York: Modern Library, 2002.

Miller, Courtney. Note, "Reflections on Protecting Conscience for Health Care Providers: A Call for More Inclusive Statutory Protections in Light of Constitutional Considerations." *S. Cal. Rev. L. & Soc. Just.* 15 (2006): 327.

Miller, Jed. Note, "The Unconscionability of Conscience Clauses: Pharmacists and Women's Access to Contraception." *Health Matrix* 16 (2006): 237.

Minow, Martha. "On Being a Religious Professional: The Religious Turn in Professional Ethics." *U. Pa. L. Rev.* 150 (2001): 661.

Mirvis, David M. "Physicians' Autonomy—The Relation between Public and Professional Expectations." *New. Eng. J. Med.* 328, no. 18 (1993): 1346.

Mitka, Mike. "Looming Shortage of Physicians Raises Concerns about Access to Care." *J. Am. Med. Ass'n* 297, no. 10 (2007): 1045.

Monroe, James A. "Morality, Politics, and Health Policy." In David Mechanic, Lynn B. Rogut, David C. Colby, and James R. Knickman, eds., *Policy Changes in Modern Health Care*. New Brunswick, NJ: Rutgers University Press, 2004.

Moreno, Sylvia. "Case Puts Texas Futile-Treatment Law under a Microscope." *Washington Post*, April 11, 2007, A3.

Morrison, Jill C. "Dr. No? The Debate on Conscience in Health Care." Panel Discussion Hosted by the Pew Forum on Religion and Public Life, the American Constitution Society, and the Federalist Society, September 8, 2006. http:// pewforum.org.

Mueller, Stephanie, and Dudley, Susan. "Access to Abortion." National Abortion Federation. 2003. www.prochoice.org.

Mundy, Liza. "Too Much to Carry." *Washington Post*, May 20, 2007, W14.

Murphy, Sean. "Freedom of Conscience and the Needs of the Patient." Presentation at the 2001 Obstetrics and Gynecology Conference: New Developments– New Boundaries, Banff, Alberta. www.consciencelaws.org.

Murphy, Sean. "Service or Servitude: Reflections on Freedom of Conscience for Health Care Workers." www.consciencelaws.org.

Murray, Elizabeth, Lo, Bernard, Pollack, Lance, Donelan, Karen, and Lee, Ken. "Direct-to-Consumer Advertising: Physicians' Views of Its Effects on Quality of Care and the Doctor-Patient Relationship." *J. Amer. Bd. Fam. Practice* 16 (2003): 513.

National Abortion Federation. "Fact Sheet on Certified Nurse-Midwives, Nurse Practitioners, and Physician Assistants as Abortion Providers." 2005. www .prochoice.org.

National Women's Law Center. "Pharmacy Refusals: State Laws, Regulations, and Policies." 2006. www.nwlc.org.

Nelson, Justin B. "Antitrust: Relevant Market Analysis of Health Care Providers." *Health L.* 9 (1997): 6.

Nelson, Lawrence J., and Nelson, Robert M. "Ethics and the Provision of Futile, Harmful, or Burdensome Treatment to Children." *Critical Care Med.* 20, no. 3 (1992): 427.

Nowlan, Mary Hegarty. "Women Doctors, Their Ranks Growing, Transform Medicine." *Boston Globe*, October 2, 2006. www.boston.com.

O'Brien-Pallas, Linda, Baumann, Andrea, Donner, Gail, Murphy, Gail Tomblin, Lochhaas-Gerlach, Jacquelyn, and Luba, Marcia. "Forecasting Models for Human Resources in Health Care." *J. Advanced Nursing* 33, no. 1 (2001): 120.

O'Callaghan, Nora. "Lessons from Pharaoh and the Hebrew Midwives: Conscientious Objection to State Mandates as a Free Exercise Right." *Creighton L. Rev.* 39 (2006): 561.

Oddens, Björn J. "Women's Satisfaction with Birth Control: A Population Survey of Physical and Psychological Effects of Oral Contraceptives, Intrauterine Devices, Condoms, Natural Family Planning, and Sterilization among 1466 Women." *Contraception* 59, no. 5 (1999): 277.

Ofgang, Kenneth. "State Supreme Court Agrees to Hear Lesbian Insemination Case." *Metropolitan News-Enterprise*, June 15, 2006. www.metnews.com.

Onishi, Norimitsu. "In Japan's Rural Areas, Remote Obstetrics Fills the Gap." *New York Times*, April 8, 2007.

Orentlicher, David. "Trends in Health Care Decisionmaking: The Limits of Legislation." *Md. L. Rev.* 53 (1994): 1255.

Ostrom, Carol M. "Prescriptions Must Be Filled under Newly Adopted Rule." *Seattle Times*, September 1, 2006. http://seattletimes.nwsource.com.

Ozar, David T. "Profession and Professional Ethics." In Stephen G. Post, ed., *The Encyclopedia of Bioethics*, 3rd ed. New York: Macmillan Reference, 2004.

Parch, Lorie A. "2-for-1 Is No Baby Bargain, Doctors Say." *MSNBC*, June 19, 2007. www.msnbc.com.

Parsons, Talcott. *The Social System*. London: Routledge, 1951.

Pate, R. Hewitt. "Monopolization." *PLI/Corp.* 1541 (2006): 157.

Pearce, Russell G. "The Legal Profession as a Blue State: Reflections on Public Philosophy, Jurisprudence, and Legal Ethics." *Fordham L. Rev.* 75 (2007): 1339.

Pellegrino, Edmund D. "Balancing Science, Ethics and Politics: Stem Cell Research, a Paradigm Case." *J. Contemp. Health L. & Pol'y* 18 (2002): 591.

Pellegrino, Edmund D. "Character, Virtue, and Self-Interest in the Ethics of the Professions." *J. Contemp. Health L. & Pol'y* 5 (1989): 53.

Pellegrino, Edmund D. "Commentary: Value Neutrality, Moral Integrity, and the Physician." *J. L. Med. & Ethics* 28, no. 1 (2000): 78.

Pellegrino, Edmund D. "Patient and Physician Autonomy: Conflicting Rights and Obligations in the Physician-Patient Relationship." *J. Contemp. Health L. & Pol'y* 10 (1993): 47.

Pellegrino, Edmund D. "The Physician's Conscience, Conscience Clauses, and Religious Belief: A Catholic Perspective." *Fordham Urb. L.J.* 30 (2002): 221.

Pellegrino, Edmund D. "Professionalism, Profession and the Virtues of the Good Physician." *Mt. Sinai J. Med.* 69, no. 6 (2002): 378.

Pence, Gregory E. "The McCaughey Septuplets: God's Will or Human Choice?" In Helga Khuse and Peter Singer, eds., *Bioethics: An Anthology*, 2nd ed. Oxford: Blackwell, 2006.

Peres, Judy. "'Morning-After' Pill Deal Reached," *Chicago Tribune*, October 11, 2007.

Physicians for Life. "U.S. Physicians' Right of Conscience Amendment under Attack—Again." September 2005. www.physiciansforlife.org.

Pierce, Richard J., Jr., Shapiro, Sidney A., and Verkuil, Paul R. *Administrative Law and Process*, 3rd ed. New York: West Publishing, 1999.

Practice Committee of the Society for Assisted Reproductive Technology and the Practice Committee of the American Society for Reproductive Medicine. "Guidelines on Number of Embryos Transferred." *Fertility & Sterility* 86 (supp. 4) (2006): S51.

President's Commission for the Study of Ethical Problems in Medicine and Biomedical and Behavioral Research. *Making Health Care Decisions: The Ethical and Legal Implications of Informed Consent in the Patient-Practitioner Relationship*, Vol. 1. October 1982. www.bioethics.gov.

Quill, Timothy E., and Brody, Howard. "Physician Recommendations and Patient Autonomy: Finding a Balance between Physician Power and Patient Choice." *Ann. Int. Med.* 125, no. 9 (1996): 763.

Rabinowitz, Howard K., Diamond, James J., Markham, Fred W., and Hazelwood, Christina E. "A Program to Increase the Number of Family Physicians in Rural and Underserved Areas." *J. Am. Med. Ass'n* 281, no. 3 (1999): 255.

Rawls, John. *Political Liberalism*. New York: Columbia University Press, 1993.

Rhodes, Rosamond. "The Ethical Standard of Care." *Am. J. Bioethics* 6, no. 2 (2006): 76.

Rhodes, Rosamond. "The Priority of Professional Ethics over Personal Morality." February 9, 2006. BMJ Online Rapid Responses to Savulescu, Julian. "Conscientious Objection in Medicine." *Brit. Med. J.* 332 (2006): 294. www.bmj.com.

Richman, Kenneth A. "Pharmacists and the Social Contract." *Am. J. Bioethics* 7, no. 6 (2007): 15.

Rie, Michael A. "Defining the Limits of Institutional Moral Agency in Health Care: A Response to Kevin Wildes." *J. Med. Phil.* 16, no. 2 (1991): 221.

Roberts, Laura Weiss, Battaglia, John, Smithpeter, Margaret, and Epstein, Richard S. "An Office on Main Street: Health Care Dilemmas in Small Communities." *Hastings Ctr. Rep.* 29, no. 4 (1999): 28.

Robinson, Walter. "In the Line of Duty: SARS and Physician Responsibility in Epidemics." *Lahey Clinic Med. Ethics* (Spring 2006): 5.

Rosenblatt, Roger A., and Hart, L. Gary. "Physicians and Rural America." *West. J. Med.* 173, no. 5 (2000): 348.

Rothman, David J. "The Origins and Consequences of Patient Autonomy: A 25-Year Retrospective." *Health Care Analysis* 9, no. 3 (2001): 255.

Sade, Robert M. American Medical Association. "Report of the Council on Ethical and Judicial Affairs 6-A-07: Physician Objection to Treatment and Individual Patient Discrimination," referred to Committee on Amendments to Constitution and Bylaws (2007). http://www.ama-assn.org/ama1/pub/upload/mm/369/ceja_6a07.pdf.

Saha, Somnath, Taggart, Sara H., Komaromy, Miriam, and Bindman, Andrew B. "Do Patients Choose Physicians of Their Own Race?" *Health Affairs* 19, no. 4 (2000): 76.

Sanghavi, Darshak M. "Wanting Babies Like Themselves, Some Parents Choose Genetic Defects." *New York Times*, December 6, 2006.

Savulescu, Julian. "Conscientious Objection in Medicine." *Brit. Med. J.* 332 (2006): 294.

Schroeder, Steven A., and Beachler, Michael P. "Physician Shortages in Rural America." *Lancet* 345, no. 8956 (1995): 1001.

Selby, Patricia L. "On Whose Conscience? Patient Rights Disappear under Broad Protective Measures for Conscientious Objectors in Health Care." *U. Det. Mercy L. Rev.* 83 (2006): 507.

Simon, Stephanie. "Aspiring Abortion Doctors Drawn to Embattled Field." *Los Angeles Times*, May 22, 2007. www.latimes.com.

Singer, Peter. "Moral Expertise." *Analysis* 32 (1972): 115.

"67-Year-Old Spaniard Gives Birth to Twins." *CNN.com*, December 30, 2006.

Smearman, Claire A. "Drawing the Line: The Legal, Ethical, and Public Policy Implications of Refusal Clauses for Pharmacists." *Ariz. L. Rev.* 48 (2006): 469.

Smith, R. "Medicine's Core Values." *Brit. Med. J.* 309 (1994): 1247.

Sokol, Daniel K. "Virulent Epidemics and Scope of Healthcare Workers' Duty of Care." *Emerging Infectious Diseases* 12, no. 8 (2006): 1238.

Sonfield, Adam. "Rights vs. Responsibilities: Professional Standards and Provider Refusals." *The Guttmacher Report on Public Policy*. 2005. www.guttmacher .org.

Sonfield, Adam, Gold, Rachel Benson, Frost, Jennifer J., and Darroch, Jacqueline E. "U.S. Insurance Coverage of Contraceptives and the Impact of Contraceptive Coverage Mandates." *Perspectives on Sexual & Reproductive Health* 36 (2004): 72.

Sonne, James A. "Firing Thoreau: Conscience and At-Will Employment." *U. Pa. J. Lab. & Emp. L.* 9 (2007): 235.

Springer, John. "Woman Defends Decision to Give Birth at 60." *Today Show*, May 24, 2007. www.todayshow.com.

Stein, Rob. "Cervical Cancer Vaccine Gets Injected with a Social Issue." *Washington Post*, October 31, 2005, A3.

Stein, Rob. "Institute Practices Reproductive Medicine—and Catholicism." *Washington Post*, October 31, 2006, A14.

Stein, Rob. "Medical Practices Blend Health and Faith: Doctors, Patients Distance Themselves from Care They Consider Immoral." *Washington Post*, August 31, 2006, A1.

Stein, Rob. "Pharmacists' Rights at Front of New Debate." *Washington Post*, March 28, 2005.

Strasser, Mark. "The Futility of Futility? On Life, Death, and Reasoned Public Policy." *Md. L. Rev.* 57 (1998): 505.

Stull, Melissa K. "Construction and Application of Emergency Medical Treatment and Active Labor Act." 104 A.L.R. Fed. 166, § 8 (2006).

Suter, Ronald. *Are You Moral?* Lanham, MD: University Press of America, 1984.

Swartz, Martha. "Conscience Clauses or Unconscionable Clauses: Personal Beliefs versus Professional Responsibilities." *Yale J. Health Pol'y, L., & Ethics* 6 (2006): 269.

Swartz, Martha. "Health Care Providers' Rights to Refuse to Provide Treatment on the Basis of Moral or Religious Beliefs." *Health Lawyer* 19 (2006): 25.

Swick, Herbert M. "Toward a Normative Definition of Medical Professionalism." *Acad. Med.* 75, no. 6 (2000): 612.

Thavis, John. "Food, Water Must Be Provided to Vegetative Patients." *Catholic News Service*, September 14, 2007. www.catholic.org.

Thorpe, John M., and Bowes, Watson A. "Prolife Perinatologist—Paradox or Possibility?" *New Eng. J. Med.* 326, no. 18 (1992): 1217.

Toner, Robin. "Abortion Foes See Validation for New Tactic." *New York Times*, May 22, 2007.

Torda, A. "How Far Does a Doctor's 'Duty of Care' Go?" *Internal Med. J.* 35, no. 5 (2005): 295.

U.S. Census Bureau. "Income, Poverty, and Health Insurance Coverage in the United States: 2006 Report." www.census.gov.

U.S. Government Accountability Office. "Health Professional Shortage Areas: Problems Remain with Primary Care Shortage Area Designation." October 2006. www.gao.gov.

U.S. Health Resources and Services Administration, Bureau of Health Professions. "Shortage Designations." http://bhpr.hrsa.gov/shortage.

Van Bogaert, Louis-Jacques. "The Limits on Conscientious Objection to Abortion in the Developing World." *Developing World Bioethics* 2, no. 2 (2002): 131.

Varughese, Dennis. Note, "Conscience Misbranded! Introducing the Performer v. Facilitator Model for Determining the Suitability of Including Pharmacists within Conscience Clause Legislation." *Temp. L. Rev.* 79 (2006): 649.

Veatch, Robert M. "Doctor Does Not Know Best: Why in the New Century Physicians Must Stop Trying to Benefit Patients." *J. Med. & Phil.* 25, no. 6 (2000): 701.

Veatch, Robert M. "Modern vs. Contemporary Medicine: The Patient-Provider Relation in the Twenty-First Century." *Kennedy Instit. Ethics J.* 6, no. 4 (1996): 366.

Veatch, Robert M. *The Patient-Physician Relation: The Patient as Partner, Part 2.* Bloomington: Indiana University Press, 1991.

Veatch, Robert M. *A Theory of Medical Ethics.* New York: Basic Books, 1981.

Veatch, Robert M. "Who Should Control the Scope and Nature of Medical Ethics?" In Robert B. Baker, Arthur L. Caplan, Linda L. Emanuel, and Stephen R. Latham, eds., *The American Medical Ethics Revolution.* Baltimore: Johns Hopkins University Press, 1999.

Veatch, Robert M., and Spicer, Carol Mason. "Medically Futile Care: The Role of the Physician in Setting Limits." *Am. J. L. & Med.* 18, no. 1–2 (1992): 15.

Vischer, Robert K. "Conscience in Context: Pharmacist Rights and the Eroding Moral Marketplace." *Stan. L. & Pol'y Rev.* 17 (2006): 83.

Vischer, Robert K. "Heretics in the Temple of Law: The Promise and Peril of the Religious Lawyering Movement." *J.L. & Rel.* 19 (2004): 427.

Vischer, Robert K. "The Pharmacists Wars." *American Enterprise Online*, February 14, 2006. www.taemag.com.

Wadleigh, John. "Clinical Case: Should I Stay or Should I Go? The Physician in a Time of Crisis." *AMA Virtual Mentor* 8 (2006): 208.

Waldron, Jeremy. *Law and Disagreement.* Oxford: Oxford University Press, 1999.

Wang, Fahui, and Luo, Wei. "Assessing Spatial and Nonspatial Factors for Healthcare Access: Towards an Integrated Approach to Defining Health Professional Shortage Areas." *Health & Place* 11, no. 2 (2005): 131.

Wardle, Lynn D. "Protecting the Rights of Conscience of Health Care Providers." *J. Legal Med.* 14 (1993): 177.

Washington Board of Pharmacy. "Rule Summary and Compliance Document." https://fortress.wa.gov/doh.

Wear, Stephen, Lagaipa, Susan, and Logue, Gerald. "Toleration of Moral Diversity and the Conscientious Refusal by Physicians to Withdraw Life-Sustaining Treatment." *J. Med. & Phil.* 19, no. 2 (1994): 147.

Weinstein, Bruce D. "The Possibility of Ethical Expertise." *Theoretical Med. & Bioethics* 15, no. 1 (1994): 61.

Weisman, Jonathan. "Where Did the Doctor Go?" *Washington Post*, May 20, 2007, B2.

Weiss, Gregory L., and Lonnquist, Lynne E. *The Sociology of Health, Healing, and Illness*, 5th ed. Upper Saddle River, NJ: Prentice Hall, 2005.

Weissert, Carol S., and Weissert, William G. *Governing Health: The Politics of Health Policy*, 2nd ed. Baltimore: Johns Hopkins University Press, 2002.

Wendel, W. Bradley. "Institutional and Individual Justification in Legal Ethics: The Problem of Client Selection." *Hofstra L. Rev.* 34 (2006): 987.

White, Katherine A. Note, "Crisis of Conscience: Reconciling Religious Health Care Providers' Beliefs and Patient Rights." *Stan. L. Rev.* 51 (1999): 1703.

Wicclair, Mark R. "Conscientious Objection in Medicine." *Bioethics* 14, no. 3 (2000): 205.

Wicclair, Mark R. "Reasons and Health Care Professionals' Claims of Conscience." *Am. J. Bioethics* 7, no. 6 (2007): 21.

Wilder, Thom. "Pharmacy Regulation: Illinois Reaches Settlements with Pharmacists Over Dispensing of Plan B," *BNA Pharmaceutical Law & Industry Rep.* 5, no. 41 (October 19, 2007).

Wikler, Daniel. Harvard Program in Ethics and Health Presentation, April 17, 2007.

Wing, Kenneth R. "The Right to Health Care in the United States." *Ann. Health L.* 2 (1993): 161.

World Medical Association. "International Code of Medical Ethics and Declaration of Geneva." www.wma.net.

Yoder, Scott D. "The Nature of Ethical Expertise." *Hastings Ctr. Rep.* 28, no. 6 (1998): 11.

Young, Robert. "Voluntary Euthanasia." In Edward N. Zalta, ed., *The Stanford Encyclopedia of Philosophy*. Summer 2006 ed. http://plato.stanford.edu.

Zacharias, Fred C. "The Lawyer as Conscientious Objector." *Rutgers L. Rev.* 54 (2001): 191.

Index